THE GHOST LAB

Also by Matthew Hongoltz-Hetling

A Libertarian Walks Into a Bear

If It Sounds Like a Quack

THE GHOST LAB

How Bigfoot Hunters, Mediums, and Alien Enthusiasts Are Wrecking Science

MATTHEW HONGOLTZ-HETLING

PUBLICAFFAIRS

New York

Copyright © 2025 by Matthew Hongoltz-Hetling
Cover design by Chin-Yee Lai
Cover images © Dotted Yeti/Shutterstock.com; © New Africa/Shutterstock.com;
© Ohm2499/Shutterstock.com; © RistoH/Shutterstock.com;
© Vincent V2/Shutterstock.com
Cover copyright © 2025 by Hachette Book Group, Inc.

Hachette Book Group supports the right to free expression and the value of copyright. The purpose of copyright is to encourage writers and artists to produce the creative works that enrich our culture.

The scanning, uploading, and distribution of this book without permission is a theft of the author's intellectual property. If you would like permission to use material from the book (other than for review purposes), please contact permissions@hbgusa.com. Thank you for your support of the author's rights.

PublicAffairs
Hachette Book Group
1290 Avenue of the Americas, New York, NY 10104
www.publicaffairsbooks.com
@Public_Affairs

Printed in the United States of America

First Edition: May 2025

Published by PublicAffairs, an imprint of Hachette Book Group, Inc. The PublicAffairs name and logo is a registered trademark of the Hachette Book Group.

The Hachette Speakers Bureau provides a wide range of authors for speaking events. To find out more, go to hachettespeakersbureau.com or email HachetteSpeakers@hbgusa.com.

PublicAffairs books may be purchased in bulk for business, educational, or promotional use. For more information, please contact your local bookseller or the Hachette Book Group Special Markets Department at special.markets@hbgusa.com.

The publisher is not responsible for websites (or their content) that are not owned by the publisher.

Note: The location of one interaction described in this book has been altered.

Print book interior design by Sheryl Kober.

Library of Congress Catalog Number: 2024040647

ISBNs: 9781541703971 (hardcover), 9781541703995 (ebook)

LSC-C

Printing 1, 2025

For Heron.

When I began this book, you were too small to see. By the time I finished, you were larger than life. Though I love you with all my heart, I cannot match the fullness of your love for beetles, sewer grates, fluttering leaves, sloths, letters, butterflies, twinkle lights, baths, worms, numbers, and—especially and always— mud. May you love life so wholly for all your days.

CONTENTS

Introduction: A Solid Phantom   1
Prologue   9
One: An Era of Spirits   13
Two: You Better Handcuff Me   21
Three: The Mantis Lay Dying   25
Four: Colleges Aren't Catapults   33
Five: Bones and Sugar   39
Six: Space Between Lives   47
Seven: Powerful Cosmic Fluids   53
Eight: More Bite   61
Nine: Very Bad Bedfellows   69
Ten: What Is the Soul?   77
Eleven: The Spiritualist Wind   87
Twelve: In a Suit of Armor on Roller Skates   97
Thirteen: Smoke Break. Smoke Break.   107
Fourteen: A Whisper in Her Ear   117
Fifteen: Our Etheric Bodies   127
Sixteen: It Sweeps Your Doubts Aside   135
Seventeen: Are They Getting Fresh?   145
Eighteen: A Crane, a Lemur, an Ocelot   155
Nineteen: The Ouija Board   167
Twenty: You Realize Lex Luthor Loses   173
Twenty-One: For That I Am Truly Sorry   183
Twenty-Two: The Force Majeure   195
Twenty-Three: These Weren't Human Doctors   205
Twenty-Four: The Scrying Mirror Hung   213
Twenty-Five: A Stake in the Chaos   221
Twenty-Six: You Freaking Dickhead   229
Twenty-Seven: She Just Started Doing Stuff Without Asking   241

Contents

Twenty-Eight: For My Highest Good	251
Twenty-Nine: That's Not Intuition	259
Thirty: Consider the Jinn	267
Thirty-One: Humor and Frequent Love-Making!	275
Thirty-Two: The Unexpected Passing of Michael	285
Thirty-Three: Almost Certainly Completely False	291
Epilogue	299
Acknowledgments	*307*
Bibliography	*309*
Index	*325*

Introduction

A SOLID PHANTOM

Its deeper tones sound the depth of the past; its more thrilling notes express an awakening to the infinite, and ask a thousand questions of the spirits that are to unfold our destinies, too far-reaching to be clothed in words. Who does not feel the sway of such a voice?
—Margaret Fuller, *Woman in the Nineteenth Century*

Do you believe in ghosts?

I do. Occasionally.

During my childhood in a depressed New England mill town, I woke up one night to find a ghost at my bedside. It wasn't in my field of view, but I sensed its menacing presence between the foot of my bed and a low set of windows overlooking the garage. I called out to my family for help, my voice thin with fright. No response.

I decided that I simply had to look at the ghost. When it came for me, as seemed inevitable, how else would I know in which direction to recoil? But the instant I turned to confront the spirit, something hard struck me in the head. It felt as if I'd been harpooned in the temple. A lion would have roared, but I'm a possum. I froze. The spirit, I thought, did not want me to move.

And so I lay, whimpering in the dark. As the minutes dragged by, my terror dulled into a resigned misery, and then a dejected boredom. Somehow, I fell back asleep.

By morning's light, I discovered that the cord of my bedside lamp had ridden up from its usual spot between mattress and headboard. As I turned to see the ghost, I'd tugged the cord, causing the lamp to fall from the narrow shelf above my headboard and strike me on the head.

In the span of hours, I had touched two worlds. In one, a ghost threw a lamp at my head. In the other, the lamp came down by virtue of the mundane weight of my mundane noggin.

Perhaps you have spent time in both worlds as well.

America certainly has.

Ghosts came back to me in an unexpected way when I began to explore the rapid ebbing of institutional trust in America, which has caused experts and national leaders to get increasingly handwringy. Pew research polls show that, between 2001 and 2021, trust in government fell from 52 percent to 24 percent. That 24 percent figure becomes even more shocking when you consider that roughly 15 percent of the workforce is employed by the government.

This trend raises troubling questions. Why is trust breaking down? Can governmental institutions even function for people who don't seem to care about them? And, if those institutions break down entirely, what sort of society lies on the other side?

To that last question, at least, I have an answer: New Hampshire.

Buried in the national data are numbers showing that the Live Free or Die state has been an early leader in the distrust business.

Trust is "the basic ingredient of social capital," according to the Carsey School of Public Policy at the University of New Hampshire, which was founded in 1893 with a mission of elevating the Granite State's public through education. As part of that mission, the Carsey School completes regular surveys to track who holds the people's trust.

The results show that New Hampshire is on the leading edge of a slow-breaking tidal wave of distrust.

A Solid Phantom

In 2001, when a majority of Americans trusted the government, just 30 percent of New Hampshirites said the same. And by 2019, the number stood at a marginal 14 percent. Or put another way, those who trusted their government were equal in number to those who thought government was controlled by Satan-worshiping pedophiles, which polling suggested was also at 14 percent.

Granite Staters also indicated that they no longer trusted their local government. Or their local media. Or even their neighbors.

New Hampshire's distrust of social institutions cut across all demographics. The men and the women, the PhDs and the GEDs, the boomers and the Gen Xers, the religious and the impious—all were united under a banner of suspicion.

One problem, as the Carsey School noted, is that these levels of mistrust not only impede institutions, but also hurt the people who distrust them. Research shows that institutional distrusters are more likely to be insecure, and therefore are less likely to build connections with other people. And indeed, New Hampshire residents living under this pall of dubiety are among the least likely in the nation to do favors for neighbors, connect with people, or help friends or extended family with food, housing, or money.

But surely, the public trusted *someone*. Even the most paranoid among us goes home and divulges their innermost thoughts—to the spouse, and if not to the spouse, then to the kids, and if not to the kids, then to the family beagle, which reliably nods along dolefully at theories on the connection between global elites and the price of potatoes.

Who is the public's beagle? And when the time comes, will our chosen confidant nod along and lick our hand? Or will it sink its teeth so deeply into our calves that we will realize we've made a terrible mistake?

As I began interviewing New Hampshire residents, it became clear to me that their granite-flecked hearts did in fact hold trust. Their trust was being extended to a group that does not show up on surveys by the Carsey School or, indeed, almost anywhere else: paranormal enthusiasts, who often identified each other, appropriately, as "believers," a term that embodies the idea of trust.

Paranormal believers offered new answers for old questions, in a sweeping variety of fields. New Hampshirites who heard disembodied voices and footsteps in their homes were seeking out ghost hunters over psychiatrists and law enforcement. Questions about the afterlife were directed to a medium rather than a minister. Those seeking wealth or career guidance eschewed financial planners and career counselors in favor of manifesting the future they desired. And addicts turned to extraterrestrial experts, rather than social services, to explain their lost time.

To understand this movement and its dynamics, I spent roughly two years studying the Kitt Research Initiative, a small nonprofit paranormal organization in the New Hampshire Seacoast region.

KRI's members, like thousands of groups across America, are driven by a tantalizing quest for the cryptids living in our woods and waters, the aliens soaring through our skies, the spirits whispering into our ears. They seek a Solid Phantom—a key piece of evidence that turns a myth into a fact, and simultaneously applies a wrecking ball to conventional thought.

Solid Phantoms do turn up, from time to time.

In 1675, Dutch clothing merchant Antonie van Leeuwenhoek peered through a self-made microscope and ended centuries of speculation about the causes of disease by making the first observation of what he called animalcules—germs and other microorganisms.

Another Solid Phantom came to light in 1861, when explorer Paul Du Chaillu strode into the hallowed halls of London's Royal Geographical Society with a stuffed gorilla in tow, thereby proving the existence of a mythical man-ape.

And a Solid Phantom was found in 1971, when astronomers at the Netherlands' Leiden Observatory captured signals from Cygnus X-1, a celestial object smaller than Oklahoma but twenty-one times more massive than the Sun—the first observed black hole, once considered a flaw in the theoretical models underpinning astrophysics, rather than an actual thing.

The Solid Phantom is not an anecdote about a mysterious glimpsed shadow or a personal experience of spirit voices. It is something tangible that can either be stuffed and put on display in a museum, or be consistently observed under controlled conditions by various independent parties. It is an explanation, or the start of one.

Given scant attention by power structures that hold their premise in disdain, KRI and thousands of groups like them are largely, like the spirits they study, invisible, free to roam America's darkest corners in a ceaseless search for their own versions of Solid Phantoms.

It is this endless journey that is subtly reshaping the country away from a set of centralized, science- and church-based homogenous standards, into a pastiche of provincialism and myth. To understand the ghost hunters and their paranormal compatriots is to understand America's near-term future.

During KRI members' journeys into the unknown, some people died, and others were reborn. They danced with aliens in the sky, and were possessed by malevolent spirits. They shared chuckles, and tears, as they tried to grapple with a different understanding of the world, and of each other.

A great many times in this book, you may find yourself wondering whether the paranormal experiences they describe are real. After hundreds of hours of interviews and observations, I have thoughts. By and large, the people I spoke with are not lying, not crazy, and (mostly) not stupid. I report on their experiences credulously—if someone tells me that they heard a voice, or saw an apparition, I write that they heard a voice, or saw an apparition. Ultimately, I am less concerned with the validity of their ghostly beliefs, and more concerned with how those beliefs are affecting their lives, and what this means for society as a whole.

At the same time, it's difficult not to get tangled up in the question of truth, and so I also include information that a skeptic would offer as evidence of bullshit. You can draw your own conclusions.

In 2022, I first met KRI's founder, Andy Kitt, in New Hampshire's Seacoast. Anchored by Portsmouth, the region is surrounded on three sides by water—the Atlantic Ocean proper, the Piscataqua River, and a massive natural harbor known as Great Bay. The overall effect is as if the Atlantic Ocean reached in and hugged this bit of land to itself to honor a kinship based on chaos and stochasticity.

Introduction

The heavily developed Seacoast attracts tourists with its aggressively performative take on historical maritime culture: fake pirates with plastic parrots perched atop their shoulders, brightly colored houses along the waterfront, seaside arcades, beachfront motels, the Seacoast Pirates youth baseball team, seashell-based artworks, and dockside lobster pounds where the soup, the salad, the mac and cheese, and sometimes even the ice cream feature lobster.

It's like a tourist roadmap that's been jammed onto a bed of rusty nails—markers of entirely different eras jut up here and there. There are massive mills and factories slowly submitting to the salty air, a shipping port from the 1970s, homes that were built in the 1600s, a military air base from the 1930s, sewer systems made of century-old bricks, cobblestoned streets, and the sprawling Portsmouth Naval Shipyard, built in 1800 and now decommissioned. And those nails are rooted in the land itself—the glacial deposits of granite that gave the state its nickname, the wind on the mountaintops, and the lapping of the ocean.

These rusting, rotting structures are reminders of the great institutional beasts whose times have passed. And now, we live in thrall to a new set of institutions—government, the church, corporate monoliths, colleges, and hospitals. Their time, too, may be nearing an end.

This was the backdrop against which I first had conversations with Andy Kitt.

Andy is proudly outspoken; he sees himself as a promoter of truth and decency, and is very willing to bruise egos and feelings along the way. He derides all things politically correct, and feels that the desire to be inoffensive is a great social ill. His ability to engage me on topics that I didn't care about could fluster me.

Andy is easy to like, what with his boundless confidence, his willingness to engage in straight talk, the way he always seems ready to hold forth on any of a wide range of eclectic topics. Also, Andy is easy to hate, what with his boundless confidence, his willingness to engage in straight talk, the way he always seems ready to hold forth on any of a wide range of eclectic topics. Andy is philosophical about his tendency to rub people the wrong way.

"It is far more common," he opined, "for someone to hate me, than for me to hate them."

Certainly, he knows a lot about a lot, and he often impressed me. When I asked him about a project he was working on that involved human vision, his response was so laden with scientific jargon that I couldn't really follow.

"There are thin lines versus fat lines and lots of intermediate lines," he said. "Each kind of line has a different threshold. I have to come up with a universal way of establishing a log scale based on the threshold—the expected threshold—that means I will never hit the diplopia. I want ten steps between no disparity and max disparity. I want the threshold to be about step three or step four. Probably step three. I'm trying to come up with a mathematical way of doing this that's not gonna interpret a one as a zero. Or two as negative one one point something. Those are the things I'm trying to do."

He also spoke knowledgeably about such diverse topics as the specific strategic military decisions of the Russo-Ukrainian war, the significance and origins of the Teenage Mutant Ninja Turtles franchise, and how to make the perfect snickerdoodle cookie. He can do woodwork. He can program a computer. He'd scored a respectable 1250 on his SATs, while drunk. He was a proud member of IQ-based societies like Mensa and Intertel.

But at the same time, the simplest tasks could befuddle him.

Once, as we walked into a Chinese restaurant in Exeter, I'd watched him futilely push against the door marked with a "Pull" sign. After the meal, when the server brought us takeout containers, she looked on in mute consternation as he piled his leftovers onto the shallow top of the Styrofoam clamshell.

Though his no-nonsense, short-cropped brown hair and glasses carry a level of authority, Andy is not a tall person. In his discussions with me, Andy frequently mentioned height—both his own and that of the people he'd known. Andy told me that his father, Peter Kitt, had been 5′6″ when he was in the military. But by the end, age had reduced him down to a saucy 5′2″.

As a seasoned journalist, I recognized that this gave me an opportunity to broach a potentially sensitive subject. I knew his father, Peter Kitt, had died in September of 2007. I asked Andy, gently, how he had died.

"Five six," said Andy. "That's shorter than average, but not by a lot."

"Yeah," I said.

"It's one standard deviation," said Andy.

"Yeah," I said. I was thinking about his dead father. "Yeah, yeah."

"You know," said Andy. "The funny thing is that the average height of males right now is declining. It was five ten. Now it's five nine."

That was interesting.

"Really?" I asked. "Why is that?"

"Because apparently," Andy confided, "all the women waiting for six footers aren't finding them." He laughed. "They're finding out the six, the six footers aren't actually gonna, like, marry 'em."

I considered this. Did it make sense that tall people were reproducing at lower rates than short people?

"What—" I started to ask. Then I caught myself.

"Uh. How. How did, uh—how did your dad die?"

Prologue

To be, or not...

—William Shakespeare, *Hamlet*

It was near midnight on Friday, and Peter Kitt knew he might die soon. The stark, silent specter of death had stalked him for weeks now, ever since he'd been admitted to the Community General Hospital in Syracuse with late-stage small cell lung cancer, a notoriously fast-moving killer associated with smoking.

Peter had been going in and out of lucidity. When he was out of it, he saw people and places long gone from his life. He saw his mother, a Ukrainian American who had lost her first husband to the Spanish flu.

He told his wife, Betty, about the vision. Had his mind simply conjured an image, or was it a ghost, somehow carrying a message from another world?

Carrying messages. Peter knew something about that. He was president of the amateur radio club in the Syracuse area. Long before that, during World War II, he'd been a staff sergeant in the Army Signal Corps, tasked with conveying information to and from the battlefield. They had overcome seemingly impossible distances by using emerging technologies—clunky, first-generation shortwave radios, multichannel FM relay sets, walkie-talkies, pulsed radar systems, and backpack radios worn by infantry. Peter and his contemporaries spent months hunched over these strange electronic systems that plucked sounds out of thin air, threw them invisible and inaudible across vast ranges, and then reconstituted them into something resembling their original form.

Was that what ghosts were? Messages from beyond?

Peter doubted it. He'd said as much to his adult son, Andy, during a Christmas visit.

"When you're dead, you're dead," Peter had said. "Life is like a switch. When you turn it off, it stays off." It was a hard truth, but hard truths were Peter's brand, delivered in a deep, gravelly voice reminiscent of Charles Bronson.

And now, at eighty-three, it looked like the time had come for him to be turned off. The elevation of the bed let him see that his bodily fluids had collected in his feet, swelling them up like worn rubber tires.

Tires.

He owned a garage, Pete Kitt's Service, in Camillus, a small town bordering Syracuse in upstate New York. Locals recognized his fleet of tow trucks because they were a distinctive shade of purple. Peter had ordered dark-blue paint, but the company screwed up the order, added too much red. The resultant cheery magenta belied his conservative outlook on life. But it was good for business.

Perhaps Peter would beat death, somehow. His heart had taken a real beating over the years. He'd undergone a double bypass, and the installation of a pacemaker. Once, in 2001, Betty had walked into the living room and found him in his chair. His heart had stopped. When she called 911, they transferred her to the nearest EMT, who only lived a block away. The EMT sprinted toward the Kitt home in his pajamas, while Betty stayed on with the dispatcher, who asked if she knew CPR.

She did not.

The dispatcher said that it would be difficult to administer CPR to Peter in his chair. "Do what you can to get him laying flat on his back," the dispatcher said.

Betty grabbed his legs and yanked. When Peter's torso hit the floor with a bump, his heart restarted, moments before the panting EMT burst in the door.

The scare had tuned Andy and Peter's other children in to his needs and thoughts like never before, which was sometimes more difficult for

Peter, who much preferred the strong-but-silent role of quiet observer. They pried details out of him. What did he want done when he died?

He wanted to die alone, he said.

Now, that seemed most unlikely. Andy and the rest of the kids were flowing in and out of his hospital room almost constantly. Even at night, hospital staff kept checking in on him, and Betty rarely left his side.

During another deathbed vision, he insisted to his visiting family members that somebody was building a bridge in the hospital room. Help build it, he'd demanded. They humored him, pantomiming the grabbing and piling of bricks into an imaginary pathway to—what? Peter didn't say.

"No, no!" he barked at one of the kids, who froze, his hand holding an invisible brick in midair. "*You* have to sit *there*, and run the drill."

Andy was the fifth of six kids. Peter watched him grow up with a mixture of admiration and exasperation. Andy was short, and overweight. He was very bright, but so socially awkward that he would stand in front of the mirror, rehearsing how to say the word "Hello" in an unrehearsed way. "Hello!" Andy said it over and over again, for hours. "Hello. Hello!"

Andy looked up to his father. When Peter allowed the kids to join him in his hobby of stamp collecting, Andy seized the opportunity. Each Sunday afternoon, Peter spent an hour reviewing his stamp albums from the living-room easy chair. Andy would lie on his stomach on the floor, paging through his own album and a stack of annual catalogs to identify particular stamps. Even here, they rarely spoke to one another. But to Andy, the time with his father was a treasure.

Among the kids, Andy in particular struggled to find space within his father's shadow. He careened between periods of overachievement and blatant rule-breaking. Andy worked at his father's garage and achieved nearly straight As through tenth grade. But then, to avoid the stereotype of a nerdy music geek, he quit playing violin in the school orchestra and took up smoking at fourteen. As a high-school senior, Andy entered a punk-rock phase and began skipping school. That year, Andy fell asleep at the wheel of his '75 Chevy Caprice while heading to a party, injuring himself and two friends.

Andy made it to college. During an early break, he came home with a silver-dollar-sized hickey, which he tried unsuccessfully to hide under a turtleneck. Betty did not like meeting Andy's paramour via hickey. "What, is that your mark of manhood?" she asked him derisively.

At five minutes before midnight, the stillness of the hospital was pierced by a fire alarm. Betty milled out onto the pavement with a night crew of staff and visitors. They filed back in seven minutes later.

Peter Kitt had, per his wishes, died alone.

He'd lived a long and full and rewarding life. And if that meant nothing, then perhaps nothing meant anything.

One

AN ERA OF SPIRITS

*New Hampshire raises the Connecticut
In a trout hatchery near Canada,
But soon divides the river with Vermont.
Both are delightful states for their absurdly
Small towns—Lost Nation, Bungey, Muddy Boo,
Poplin, Still Corners (so called not because
The place is silent all day long, nor yet
Because it boasts a whisky still—because
It set out once to be a city and still
Is only corners, cross-roads in a wood).*
—Robert Frost, "New Hampshire"

When Peter Kitt died, his story did not end.

A night or two after his passing, Betty's cellphone disappeared from her night table while she slept. She asked Andy (who was staying in the family home) and the other adult kids to help find it. When they called the number, they found it was sitting atop Peter's obituary picture.

This was the first of hundreds of experiences over the next several weeks. Though they took the usual seasonal steps to warm the house as the fall of 2007 yielded to winter, they found doors and windows open to the freezing weather. Peter's alarm clock went off between two and four a.m., nearly every night, even when the alarm function had been turned off. Appliances broke down. The phone line went dead. A serviceman came to repair it, and said the lines to the house and to Peter's study had been switched around, which no one had done. A few days after he fixed it, it went dead again. A second serviceman also said the lines had been switched around. Another time, they found that Betty's cellphone had somehow swapped places with Peter's cellphone, which had been packed into a box of his personal effects in a closet.

It was a near-constant barrage of low-key mysteries. Any one of them could be explained away, Andy thought, but all of them, together? He was by then in his late forties. He had never really cared about the paranormal, but he now began to wonder if his father's spirit was in the home, trying to communicate.

Andy wanted what so many bereaved do—contact with his departed loved one. But was his father's spirit accessible? Or were all these events in his mother's home a confluence of coincidence and wishful thinking?

When he returned to his own home in the New Hampshire Seacoast, Andy began reading everything he could on the supernatural. He consumed dry and technical academic studies, thrilling but not-credible stories from those who claimed to have contact with the dead, and everything in between. Some said ghosts were responsible for all hauntings; others saw evidence of demons, other nonhuman entities, and even extraterrestrials. He read theories of ghost sightings as unintelligent echoes of the past, and of ghosts as sentient beings from the future, from parallel universes, or from a timeless place that allowed them to experience Earth's past, present, and future all at once.

In what he would later think of as his first experiment, Andy spoke into a handheld recorder, addressing his remarks to his father. He described his own theory, how he perceived the post-death experience—a description that involved souls, and different time perceptions, and a kind of purgatory.

With the recorder running, Andy asked his question: Was he correct?

When he played the recorder back, he heard the unmistakable Bronson-esque voice distinctly. It was only a single word.

"Yes."

But it was enough.

"He answered yes, with absolute clarity," Andy said. "It's not—it wasn't a question, wasn't like, 'Yeah, maybe.' It was 'Yes.'"

After high school, Andy lurched from enterprise to enterprise, never quite settling into the sort of permanency that marked the typical American life. After a military post in Alaska, he moved to the state with his wife and daughter and found a field job with the census. He maintained a side hustle of working a gold claim that barely paid for itself. In 2000, he and his wife split up, and he moved to New Hampshire, into the home of a woman with whom he'd struck up a long-distance romance. It didn't work out.

"The first night, we were sleeping together," he later said. "The third night I was trying to explain to her husband, the trucker, who I was."

Andy had continued to collect stamps, amassing a trove that, to his everlasting pride, vastly outstripped that of his father. He opened a comic-book store in the 1990s, and then closed it with spectacularly bad timing, just before the internet and the Marvel Cinematic Universe sent values skyrocketing. He took odd jobs, and started applying to temp agencies. If it was office work, his hunt-and-peck method got the job done. If it was manual labor, he did what he could to keep up with men who were larger and more fit.

His brains and his deep bench of skills kept him going—he bought a sewing machine so he could tailor three-dollar pants from the Salvation Army, and often bought beater cars at auction, drawing on his mechanical skills to fix them up. But by the time his father died, he was in his late forties, taking classes at the University of New Hampshire, and frazzled by the constant strain of poverty. He was living in a former turkey coop that he had rented on the shores of the Seacoast's Great Bay. He made the rent by selling kayaks, tapping student loans, and setting up booths at trade shows to sell *pysanky*—eggs that he intricately decorated with wax and dye, in the tradition of his Ukrainian ancestors.

In late 2007, after Andy heard his father's voice speaking to him from beyond, he undertook his own quest for a Solid Phantom, one that would prove that his experiences with his father were real.

Using the online platform Meetup, Andy announced the first meeting of the Seacoast Paranormal Research Group. He reserved a table at a Dover bar, Central Wave, and waited to see who showed up. Two people came. At the next meeting, in November, one person showed up.

In December, Andy announced via Meetup that he'd had to cancel the meeting. This was a lie. The truth was, no one had come. He'd eaten a lonely meal and gone home, unsure of whether SPRG existed in reality, or only in his mind.

SPRG's failure to launch was surprising, because in 2007, a belief in ghosts was ascendant. Polls showed that 32 percent of Americans believed in spirits, triple the share that had believed in the late seventies. And people didn't just believe. Millions more Americans than ever before were reporting personal experiences with ghosts. Tens of thousands were suddenly brushing up on methods of paranormal investigation so that they could roam old buildings with recording devices and severe cases of the heebie-jeebies.

Certainly, the ghosts must have noticed that something was going on. For decades, they had been left in relative disembodied peace, and now a small army of the living was suddenly prying into every cobwebbed corner of every ramshackle structure and using a mix of tech-based gadgets and psychic powers to establish a communicative rapport.

What to the dead must have seemed like an unprecedented era of organic intrusion seemed, from the perspective of the living, to be an era of spirits. And interestingly, more than spirits. There were increases in reports of other strange things—UFOs and their alien pilots, Bigfoot and other cryptozoological marvels, witch spells, supernatural entities that had never been human, spirit guides and demons and angels and evidence of the lost city of Atlantis.

In January 2008, Andy was surprised when his SPRG meeting drew a host of unfamiliar faces. Thirteen people had shown up, almost all first-timers. With barely suppressed excitement, Andy asked them to go around the table and introduce themselves.

By the time introductions were halfway through, Andy's feeling of enthusiasm was already yielding to one of contempt for his fellow truth seekers, whom he later described, in his characteristically brusque manner, as "fluffy donuts."

"Half of them, I was like, 'Wow, you're just freaking crazy,'" Andy said. Each person claimed to be in contact with the dead, and each asserted a level of mastery and control that rang false to Andy's ears. His crest fell a bit more with every introduction, until it got to a woman in her mid-thirties with thick, long brown hair and a khaki-friendly fashion sense. She alone among the attendees seemed to be experiencing the event with Andy's level of cringe.

She gave her name—Isabeau Esby. "I'm a medium," she muttered, palpably eager for the spotlight to move on so that she could slink out the door unnoticed.

Andy's interest was immediately piqued.

Isabeau "Beau" Esby doesn't like to describe her early years. "I had the type of childhood that you sort of learn not to remember," she said, which has made her habitually forgetful, particularly about her own interactions with spirits.

At twenty-two, she graduated Minnesota State University with a major in mathematics and a minor in chemistry. That same year, 1996, she married Troy Esby and moved to New Hampshire, after which she juggled raising two children, a cocker spaniel named Apple, and a series of family business enterprises. She ran her own accounting firm. She and Troy created a flooring company, where he did the labor and she handled much of the administrative and leadership work. In 2004, she opened the doors for the Stratham Fitness Center and Holistic Spa, which offered everything from nutritional counseling to aromatherapy. She was a typical granola-leaning but entirely mainstream mom. She liked yoga, movies, and tea.

But in 2005, when Beau was thirty-one, she entered her living room one day to find her beloved grandmother, who lived hundreds of miles away and had no reason to be there. Particularly since, as Beau had just learned, she'd died a few hours ago.

The spirit of Beau's grandmother returned, not once, but many times. She spoke to Beau with as much clarity and substance as if she had still been alive. She said Beau had a role to play in the spirit world, and introduced her to Walking Elk, an intimidating Indigenous spirit guide. They told Beau that her life path was to become a hollow bone, an instrument through which spirits could communicate messages that the living needed to hear.

Other spirits began to appear to her. They were trapped in a limbo, unable to make a clean break from their mortal existence. Whenever possible, she helped them to move on.

The spirits typically appeared just at the edge of her vision, shimmering like heat. Beau compared it to a bowl of strawberries embedded in an ethereal Jell-O.

"If you stop looking at the strawberry and kind of relax your vision, you can see the Jell-O," she said.

When I first met Beau, I was instantly drawn to her quick humor and tendency to clown around. She has a sort of easy wisdom to her, and a way of ever-so-subtly deflating tension in any situation. For example, when she was recounting a memory, she didn't say, "He screamed in panic. It was bone-chilling."

No, she said, "When you heard him, like, scream in panic, it was kind of bone-chilling, man."

Who produced the panicked scream, and thereby induced Beau's chilled bones? We'll come back to that.

When she began speaking to spirits for money in her living room, others were also drawn to her. Fueled mostly by word-of-mouth, almost overnight, Beau had a level of commercial success that few mediums ever attain. She would be booked out months in advance.

But the stampede of paying clients only left her troubled. They wanted answers to their questions, but her guides told her that her true purpose was helping the spirits, who often wanted to impart information that the living did not want to hear.

Beau longed for a different dynamic. She wanted to play in the spiritual sandbox. It occurred to her that she might find what she was looking for through one of the groups of ghost hunters that was popping up in New Hampshire.

"I didn't want to ghost hunt," said Beau. "I wanted to be in a state where I could read without a paying client seeking assistance and help."

One local group took her to a supposedly haunted house, but she was disappointed to see that the proceedings were chaotic and performative—people shouted that they were feeling odd sensations, or stampeded toward the tiniest creak of a floorboard. They also committed the sin of "feeding the medium," meaning that they contaminated Beau's perceptions by telling her what they expected her to find.

"Do you hear a baby crying?" they asked her.

Stop, she thought. *You're not helping the situation.*

In January 2008, Beau's search for a decent ghost-hunting group led her to Andy's SPRG meeting. She agreed with Andy that the group he had founded was already overrun with kooks.

"Everyone was like, 'I'm a medium who sees horses from the third planet of something or other,'" said Beau. "They were all so out there."

But she and Andy had an easy rapport. They cracked each other up, sometimes with wit, and sometimes with crassness.

"That's my friendship vibe," she said.

Beau was also impressed by Andy's critical thinking, including a willingness to debunk. This was just what Beau needed—a forceful push toward logical expressions of her mediumship. He reeked of ambition, and projected confidence.

Before long, the two had built up a mutual respect, based in part on their agreement that the whole business of interacting with spirits should be held to a much higher standard. Andy had a vision of what that might look like, and Beau had a raw spiritual talent that could allow them to capitalize on that vision. They talked for hours about how to imbue a ghost hunt with science—not the thin veneer pasted atop paranormal reality shows, but real science.

What would a double-blind experiment look like? What equipment could hold up to scientific scrutiny? How could a medium's talent be employed in the service of this higher purpose?

"Here's somebody who's gonna get this done," Beau said to herself. "He sets the parameters. He's not wishy-washy. This was a scientific setup. And it's great."

The sky was no limit.

Two

YOU BETTER HANDCUFF ME

The universe is wider than our views of it.
   Yet we should oftener look over the tafferel of our craft, like curious passengers, and not make the voyage like stupid sailors picking oakum.

—Henry David Thoreau, *Walden*

As patrolman Daniel Frazer cruised north up Main Street in Rochester, New Hampshire, he spotted the blue 1988 Plymouth Reliant sitting in the southbound lane, engine idling. Suspicious.

One interesting thing about the Seacoast is that cops never know when any particular traffic stop is going to turn out to be alien-related. Aliens had played an outsized role for the area's law enforcement since 1965, when officers Eugene Bertrand and David Hunt, of nearby Exeter, filed reports documenting a spectacular and extended encounter with a huge, low-flying UFO in a farmer's field.

The national media had a field day with what became known as the Exeter Incident. The Exeter police force became town laughingstocks, even after the incident led then senator Gerald Ford to request a federal investigation of UFOs.

In the wake of the Exeter Incident, the region became a hotspot for UFO sightings, despite a strong stigma for the individuals who reported them. Those who said publicly that they had seen a spaceship from another planet were treated as if they had, in those less enlightened times, identified as queer, or confessed to having a mental illness. They were held up to public ridicule and mocked, and yet, the very tormentor who turned the screw would often go back home and whisper that they, or a family member, had seen a spaceship too.

The Plymouth that Officer Frazer saw idling on Main Street was blocking traffic, except that there wasn't any traffic because it was a residential area in a sleepy town at 3:44 on Sunday morning, when all but the most dedicated Saturday-night revelers had called it quits.

As Frazer drove by, he could see the headlights of the Plymouth as it suddenly jolted into forward motion. Frazer made a quick U-turn. When he saw that one of the Plymouth's rear doors was hanging open, he hit the lights. The Plymouth sped up. He followed it, and after it passed the Holy Rosary Parish, it slowed, coming to a stop in front of George & Ed's, a local general store.

Officer Frazer got out of his cruiser. The air was cold and wet, just at the freezing point. A light, meandering breeze chilled the skin.

As he approached the Plymouth, he saw why its door was open—a white sawhorse that belonged to the Rochester Department of Public Works was sticking out of the back seat. Three young men sat in the car. Frazer, who had seen such things before, many times, relaxed somewhat. Typical teenage hijinks.

The nineteen-year-old behind the wheel of the Plymouth had a quiet presence, and handed over his driver's license without complaint. By the light of his flashlight, Frazer could see this was Michael Stevens Jr., part of a family with deep ties to the area. Stevens had strong features and a solemn, steady gaze that suggested a perpetual quiet contemplation of life.

Frazer asked Stevens why he was driving with a sawhorse in the back seat. This was police protocol, a chance for suspects to explain themselves—or to entrap themselves with lies. Many people threw out a desperate verbal Hail Mary that could come back to haunt them in a police report.

You Better Handcuff Me

But Mike was no bullshitter.

"I don't really have a good reason," he said, in a deep, strong voice stamped with a rural New England accent.

Frazer asked if the sawhorse belonged to Mike. The question was, at best, a courtesy. The sawhorse had ROCH DPW painted on it in large orange letters. Other, nearly identical sawhorses were standing on a temporary construction site right there on the intersection with High Street, where they kept cars away from a depression in the road.

Again, Mike refused to take the bait.

"No," he said.

After a second officer arrived at the scene and the trio of (saw)horse thieves were cleared out of the car, things escalated. With the sawhorse leaning up against the Plymouth, Mike asked Frazer for his name and badge number. Frazer refused to tell him, and instead told Mike to turn around so he could put handcuffs on him. Mike backed away from the officers.

"You can't arrest me until you give me your name and badge number," he said.

It was a classic Moldovan standoff. Frazer refused to give him the information until Mike was handcuffed, and Mike refused to be handcuffed until he had received the information.

Frazer grabbed Mike above the left elbow. Mike jerked away. The two officers drove him back against the hood of the cruiser. Frazer tried to force Mike's right arm behind his body, to be cuffed, but Mike was a solid six feet tall, and country strong. Mike twisted until his back was to the officers, denying them access to his arms. They pulled ineffectually, but couldn't budge him.

At an impasse, Frazer reached for his pepper spray, threatening to use it to subdue Mike.

Mike considered it for just an instant.

"Fuck this," he said, and put his own hands behind his back. But when Mike was seated in the cruiser, tempers flared again. He banged his head repeatedly against the inside of the cruiser, shouted, and was generally belligerent as he was driven a few blocks to the Rochester police station, an island of brick in a sea of crumbling blacktop. While he was being processed, Mike refused to empty his pockets, and had to be cuffed again.

"You *better* handcuff me!" he shouted. "I'll . . . knock all of you motherfuckers out if I get the chance!"

They turned out his pockets—six beer-can pull tabs, plus five Budweiser bottle caps. Mike told them they didn't know who they were fucking with.

"There's always revenge, motherfucker!" he shouted. "This isn't over!"

But, really, it was over.

He would soon be charged with theft, resisting arrest, and a first offense of driving while intoxicated—all misdemeanors, which resulted in no jail time. After all, a teenager having the poor judgment to get drunk and steal a sawhorse was hardly out of place in Rochester.

There is no record of the officers asking Mike why he had taken the sawhorse, and there is no record of an answer. But the episode did have an interesting coda.

Eighteen months later, in September 1999, a different Rochester policeman went to an apartment to serve some papers on Mike for an unrelated matter. Mike had moved on, the landlord said, been gone for months. However, the landlord suggested, the officer might have an interest in seeing a couple of items that Mike had left behind.

The cop found, in the apartment, a small collection of other street signs and traffic-control devices that Mike had apparently taken from their rightful locations. Typical teenage hijinks, on steroids.

Why was Mike the kind of rambunctious kid who stole street signs, and scuffled with police? Even beyond his youth and the beer, there was an underlying reason for his belligerence, and his resistance to their restraint. As a young child, Mike had suffered severe trauma. He had been kidnapped and subjected to terrible experiences. It was so horrific that his mind had hidden most of the details from his consciousness.

The kidnappers were never brought to justice. In fact, after they returned Mike to his family, they abducted him again. And again. They were literal light-years ahead of any authorities who could bring them to justice. This is because they were extraterrestrials, using advanced technology to carry out their crimes.

This is Mike's belief. And in 1998, it was a very isolating one.

Three

THE MANTIS LAY DYING

It is easy to be brave from a safe distance.

—Aesop

Andy and Beau! Beau and Andy! As the most prominent faces of SPRG, they were suddenly everywhere in the local paranormal scene, offering a one-two punch of warm spiritualism and cold logic. They lectured in libraries and taverns, ballroom events and haunted inns. They were written up in the *Portsmouth Herald*, the *Atlantic News*, the York County Weekend edition of the *Journal Tribune*, and *Fate* magazine. One local newspaper did a long feature piece on Beau's transformation from a paranormal skeptic to an experienced psychic medium.

And every month, new people showed up to SPRG meetings. At one, a large, quiet man they had never seen before came asking for help. He had the air of rough living about him, and wore jeans, a sweatshirt, and a baseball cap.

The man was gathering petition signatures in a long-shot effort to have the New Hampshire state government erect a historical roadside marker on a scenic stretch of Route 3 in Lincoln. There, in 1961, Betty and Barney Hill became the first alien abductees to receive widespread publicity.

The elements of the Hills' experience have become staples of abduction lore—lights in the sky, diminutive humanoids with no hair and large eyes, and missing time recovered through hypnosis.

The petitioner was Mike Stevens, who had done some growing up in the decade since his own abduction of a city-owned sawhorse and subsequent scrap with the police. He'd gotten married and, in 2001, fathered a child at twenty-one. He'd done various manual-labor jobs and spent a lot of time hiking in the New England woods, both by himself and with his kids. In 2008, after surgery to address a bone spur that was shredding his Achilles tendon, he spent weeks recuperating on the couch. He passed the time with a daily diet of paranormal-themed TV shows, and a book about the Hills written by their niece, Kathleen Marden.

"Somewhere between the pain pills and all that it just kind of clicked through. Why are you watching and reading all this stuff?" Mike later said. "Why aren't you doing something about it?"

The something, for him, turned out to be the petition drive that led him to Andy and Beau. He began coming to SPRG events and meetings. His neatly trimmed coal-black beard, soulful eyes, and heavy brow made him seem larger than life; tattoos (nine of which were self-inflicted) were displayed prominently on each arm and leg. And yet, he invariably blended into the background—not a wallflower so much as a big, bushy shrub.

Andy quickly grew to respect Mike for his work ethic and attention to detail during ghost hunts. The twosome offered an almost comical set of contrasts—Mike was a hushed hulk, while Andy was short and allergic to silence. When a local newspaper reporter tagged along on an investigation, she called Andy an "artist, Mensa member, kayak salesman, ghost hunter extraordinaire . . . [and] charming conversationalist," while she described Mike as "a great bear of a fellow, and quiet as a mouse."

The buzz around SPRG—which now boasted a few dozen active members—was indicative of a buzz around ghost hunting in general in the late 2000s. And anyone who knew anything about New England's ghosts knew that the region's dominance over national ghost-hunting culture is due to two demonologists, and two plumbers.

At any given moment, hosts of fringe ideas circle the edges of our cultural mainstream, hot embers waiting for the right conditions to light a zeitgeist-warping wildfire. Some, like organic food, get there. Others, like competitive shin-kicking (a 410-year-old tradition in England's Cotswold Olimpick Games), are still awaiting their day in the sun.

Over the past century, New Englanders have caused ghost hunting to burst into flame in two distinct stages, one born of religion and one born of science.

The first was set in motion in 1943, when sixteen-year-old Lorraine Moran had a Connecticut meet-cute emblazoned with the hallmarks of the 1940s. She was at the local movie theater taking in the latest James Cagney flick and making eyes at an absolute dreamboat of an usher named Ed Warren, who worked days as a lifeguard and smelled of Noxzema. After the show, Ed walked Lorraine and a couple of her gal pals to the soda fountain across the street, and offered to buy them a round of sodas, priced at a nickel apiece. Lorraine ordered an ice-cream float, which cost a dime. When Ed didn't complain, Lorraine went home and wrote in her diary that she'd met the man she was going to marry.

As newlyweds, Ed and Lorraine developed the unusual hobby of sitting outside ramshackle houses and painting them; afterward, the Warrens would show the homeowner their work and ask for permission to come inside and search for spirits. Eventually they began battling these spirits, and calling themselves demonologists. The term was coined in the 1800s, when dropping *ology* onto any subject was something of a passion (as with horology, the study of time; sexology, the study of sex; and even Ripperology, the study of Jack the Ripper).

The demonologist Warrens believed that they were agents of God, vanquishing demons capable of wreaking major havoc, including "fires, explosions, dematerialization, teleportation, and levitation of large objects."

In 1952, the Warrens founded America's first paranormal society, the New England Society of Psychic Research, and began building a network of New England believers. They trained others in the art of demonology, spoke at college campuses, sold books, made international tours, and frequently appeared on camera, including when they took a television crew into the *Amityville Horror* house. But even at the peak of their popularity, they were famous only within the confines of their fringe interest, like a champion hot-dog eater. Or a soap-opera star. Or a libertarian nominee for US president.

Some of the Warrens' ideas exploded into the collective consciousness, such as haunted dolls, demons throwing physical objects, and slime or blood running down walls. But there is also a sizable bin of less-celebrated cast-offs. For example, they described demon-possessed people whose bodies morphed into gorillas. They also talked about being attacked on the freeway by a mysterious van that pulled in front of them and spat green goo onto their windshield. The van sped ahead, out of sight, only to come up behind them again moments later. The ghostly vehicle repeated the process, several times, in an apparent effort to make them crash.

The Warrens, thankfully, survived the gorillas and goo-shooting vans, as well as a host of other elaborate plots to kill them. Why these powerful demons, which could bat a refrigerator around like a ping-pong ball, didn't simply throw a kitchen knife into the Warrens' hearts while they slept, I can't imagine.

By the 1980s, as the Warrens aged into their sixties, their religious framing was losing relevance. Their last national television appearance was on *The Sally Jessy Raphael Show*, in 1992. As the cellphone age dawned in the early 2000s, their lack of video footage seemed increasingly incongruent with their colorful accounts.

Just when it seemed that ghost hunting would peter out into obscurity of the sort experienced by Jell-O-mold entrees and model-train aficionados, a new vision took center stage.

In 1995, Jason Hawes and Grant Wilson, a couple of plumbers who worked for Roto-Rooter in Rhode Island, started a ghost-hunting group called The Atlantic Paranormal Society, or TAPS. In stark contrast to demonologists, they preferred what was at the time a little-known term: "paranormal investigator." *Investigate*, derived from the Latin *vestigium*, means a trace, or a track. It implies a logical analysis of evidence to infer the truth of an event, and it also recast ghosts, from threatening demons to test subjects.

The Mantis Lay Dying

Hawes and Wilson were invited to headline a first-of-its-kind unscripted television show for the Sci-Fi Channel. When *Ghost Hunters* debuted in 2004, it still had one foot in the demon-battling world that had been so thoroughly established by the Warrens. The initial cast even included two demonologists, twins Carl and Keith Johnson, who had learned the craft from the Warrens themselves. But by 2006, the Johnsons were gone, and the show had fully embraced its scientific trappings. In every episode, Hawes, Wilson, and the rest of the Ghost Hunters cast used infrared cameras and electromagnetic-frequency readers to "collect evidence" for later "analysis" that provided the grist for a final "conclusion" about whether a locale was haunted.

The show was a surprise hit. Three million people tuned in to this interpretation of spirits every week, and a significant number started or joined amateur ghost-hunting groups themselves, creating a wave of public interest that buoyed the profile of Andy and SPRG.

Andy briefly met Wilson and Hawes in June 2008, during a paranormal retreat at a Holiday Inn in Batavia, New York. The highlight of the three-day event came when attendees investigated the nearby Rolling Hills Asylum, which, with its forbidding colonnades and grim history, was considered one of the most haunted places in the world. Though Andy found the ghost hunt to be chaotic and uncontrolled, he did have a powerful personal experience during a private moment on the third floor of one of the massive brick wings of the asylum. As he passed one of the rooms, he was filled with a sense of profound unease. But when he went inside, he found nothing, and filed it away as an unsolved mystery. At the end of the retreat, he was none the wiser for having rubbed elbows with the two plumbers who had sparked an era of ghosts, not as evil beings to battle, but as phenomena that could be wrestled to the ground by the scientific process.

For Andy, Mike's competence was a chunk of sweetness in a rapidly souring SPRG situation. To his extreme annoyance, no one could agree on ghost-hunting protocols, or even who should be allowed to go to any given haunted house. Throughout all of 2008, its first year of existence, hard feelings threatened to tear SPRG into factions. By early 2009, Andy decided that, if he wanted to pursue ghosts seriously, he would have to do it with a completely different group structure, one that was less inclusive, and more exclusive.

In mid-March 2009, Beau and Mike met with Andy and his girlfriend at a Portsmouth pub called Coat of Arms. The British-themed bar was dimly lit, making it difficult for patrons to judge the winners of snooker, or the visual appeal of the kidney pie.

They sat at a round, isolated table Andy liked, tucked away against a wall that afforded them privacy; Beau's back was to the open restaurant, giving her a sense of vulnerability. They agreed that SPRG, and the broader ghost-hunting landscape, was a mess. Everyone was either in it for the kicks, or a skeptic whose rejection of evidence was so knee-jerk that it made them worse than the most credulous of believers.

They had decided to found a new group, the Kitt Research Initiative, that would screen out unserious thrill-seekers and elevate the act of ghost hunting. They discussed drafting a mission statement—they wanted to validate or disprove research techniques, properly define terms like *ghost*, collect data, and act as a filter that would convey only the most promising evidence and theories to mainstream scientific institutions, and the public. Andy envisioned that the KRI seal of approval would help people separate wheat from chaff.

"Sort of like a Snopes," he said, "but it doesn't just say nothing is true."

At some point in the conversation, the focus shifted to Mike, who was, again, notable more for his silence than his speech. Mike was no stranger to ghosts. His family had always felt that his father's house was haunted. But unlike Andy and Beau, spirits were low on his priority list. He was interested in aliens, on a very personal level.

One perennial decision that alien abductees like Mike face—at this table and in his life—is knowing how much to share. It was difficult to predict the exact point at which people would go from intrigue to incredulity. Should he talk about the times that he had seen unexplained lights in the sky from a great distance? Should he reveal the tattoo of an alien on his right thigh? Should he mention the lost time, the inexplicable urges that he suspected were artifacts of alien-abduction-related trauma?

For example, while doing dishes in his second-floor apartment in Rochester, he'd opened the window so that he could lean out to have a cigarette without stinking up the kitchen. A praying mantis was resting on the window ledge. Mike grabbed a kitchen knife up off the countertop and cut it into two roughly equal halves. As the mantis lay dying, he couldn't explain to himself why he'd done it. On later reflection, he suspected that he'd responded viscerally to something in the bug that had reminded him of an alien encounter that was no longer part of his conscious memories.

He looked at the trio of attentive faces. It was difficult, sometimes, to know what to share.

Mike decided to open up, and began talking about the first abduction he could remember, which took place in 1983, when he was three or four years old at a family gathering at his grandmother's house. He and his cousin, who was about the same age, were outside the house in the evening. They'd watched a saucer-shaped craft adorned with pastel-colored lights float over the trees toward them. As they gaped, Mike experienced a blip in time; suddenly, he and his cousin were standing in a different place in the yard, and the craft was in a different position in the sky. After the blip, the UFO shot skyward, and the children went back inside.

I was not able to speak with Mike for this book, for reasons that will later become obvious, but he has made public statements about what happened next at the pub, which matches the memories of Beau and Andy.

Beau listened attentively as Mike described the blank spot in his memory, and why it seemed significant. Just as the conversation turned to the possibility of Mike trying to recover the lost time, a strange expression crossed Beau's face.

"I felt like somebody was behind me, and I also felt very uncomfortable, almost like I was being made to feel uncomfortable," she said.

The atmosphere around them turned somehow sticky and gross.

"It was a greasy, slimy feeling," said Andy. "We all felt it."

Beau was not easily shaken by her visions—she had by then communicated with a long train of the dead—but this, she said, was of a different category altogether: a being that was six or seven feet tall, and nasty.

Looking back on it a decade later, she said that her clearest memory was of the thing's hand, transparent and ghostly, reaching out from behind and into her field of vision with inhuman fingers that were perhaps ten inches long.

It ordered Beau to stop the conversation about Mike's time blip.

"He's not going to talk about it," it said to Beau. "This conversation is done."

Beau's guides had told her that, if she used her intent to protect herself, no spirit could harm her. But this felt different.

"That was one time," she said, "that I felt scared of something that was not solid."

After the creature departed and the feeling of greasiness cleared, the foursome decided to press on. Though they were shaken, the conversation returned to KRI, and its mission statement.

Mike once again receded into the background, as Mike did. But a small smile played out—if not on his lips, then internally. He glanced at Beau.

No one had ever seen the things that had been plaguing Mike.

"I finally," he said later, "felt like somebody believed me."

Four

COLLEGES AREN'T CATAPULTS

There is, I believe, in every disposition a tendency to some particular evil, a natural defect, which not even the best education can overcome.

—Jane Austen, *Pride and Prejudice*

According to one online register, nearly fifty different New Hampshire–based ghost-hunting groups popped up in 2008 and 2009. The actual number is almost certainly much higher. Spirit-hunting was so hot that it seemed the state motto could plausibly have been changed to Live Free or Die and Then After You Die Meet Me in a Dark House After Midnight and Say One or Two Indistinct Words into My Recorder.

And yet, in 2009, the state's policymakers and academics didn't seem to notice. They were more focused on a potential crisis in New Hampshire's higher-education community.

When thinking about the value of a college education, it's become standard to picture Gothic columns, domed roofs, and the sort of ivy vine coverage that would have rendered Rapunzel's hair redundant. Despite our national obsession, elite institutions like Harvard, Dartmouth, and Oxford (and *that's* how you do an Oxford comma, friends) have little to do with the purpose of college—elevating the American public to foster enlightenment and career preparedness. The entire Ivy League educates only about 1 percent of the nation's college students (160,000 out of 16 million). And before those students set foot on campus, most have already been anointed as the Wagyu steak and foam-speckled truffle fries of their generation. There is a growing sense that elite schools function largely to cement the status of the privileged families they serve, catapulting wealthy students to new levels of riches.

The best colleges aren't catapults; they're lifeboats, allowing virtually any member of the public to hoist themselves up out of troubled economic waters, and sit comfortably a few inches above sea level.

A college that serves the general public makes a difference in the most basic way. It funnels money into the pockets of its students. For every four dollars earned by a person with a high school diploma, a person with a college degree will earn seven dollars.

For visual learners, I've converted this into units of that luxurious ass-washing device, the bidet. The average high school graduate's lifetime wages allow for the purchase of a seemingly respectable hill of 1,300 bidets. But imagine the abject shame that high school graduate would feel, to see a next-door neighbor with a college degree sitting on a 2,300-bidet mountain. That's a thousand more bidets!

And those bidets allow many to live a life of quiet dignity that otherwise may not have been possible. Consider a Seacoast resident you've never heard of: Dolly Markey.

Dolly's mother Mabel, who had only an eighth-grade education, gave birth to Dolly at their home in Durham, New Hampshire, in 1928. Dolly's father, Louis, was a janitor with a sixth-grade education. Because Durham had no high school, Dolly woke up early each morning to board a train bound for nearby Dover, where she made her way to their high school. When Dolly graduated twelfth grade in 1946, she enrolled at Dover's McIntosh College, located just a mile from the high school.

Ever since it opened in 1897, McIntosh had been a college of the people, and it looked like it. The main building, with its white clapboard and blue-shuttered windows, bore more resemblance to a classic New England church or a seaside B&B than an ivory tower.

McIntosh College wasn't fussy. Anyone who filled out an application and appended a copy of their high-school transcript was accepted. It was exactly what the families of the region needed to help them get through times of economic change.

Dolly took the skills she learned at McIntosh and used them to land an office job at the University of New Hampshire, which carried a decent wage and benefits that she could not have gotten elsewhere. She was able to eventually raise four children of her own with her husband, Paul, a World War II veteran and utility worker.

It's a simple story. It's a profoundly American story. And in some ways, I admit, it's a boring story. But it's what has, for more than a century, made New Hampshire tick.

For decades, through war and peacetime, McIntosh gave otherwise underserved populations better futures. Its graduates provided the backbone that propped up utility companies, shipping docks, and, before the local mill industry faded, stitching companies. They became legal secretaries, school administrators, chefs, masseuses, computer and medical-lab technicians, and small business owners. They populated the Elk lodges and the Rotary clubs and the churches and volunteer fire departments.

During major historical events, McIntosh graduates were often present, though invariably in an underappreciated background capacity. Alumnus Leon Chetsas was among the first soldiers ordered to march toward an atomic-bomb mushroom cloud in an August 1951 training exercise in Camp Desert Rock, Nevada. Oritha Miller, another alumna, was one of the first Black women in the US Marine Corps, and served during the Korean conflict.

But somehow the communal goods provided by McIntosh slipped through the Seacoast's salt-crusted fingers. In 2004, McIntosh had roughly 1,300 students. By September 2009, it had eight. In 2010, it had zero. Nearly all of the school's equipment and resources were carted off to the town of Dover's Recreation Department, or to the Dover High School that Dolly had once taken a train to attend.

Dolly, at eighty-two, had outlived her alma mater.

If McIntosh's students were agog with the loss, those who ran in circles in which educational trends are studied were even agogger. In all of New Hampshire's history, the closure of a 113-year-old school was unprecedented.

People blamed various factors. Student applications were down. The national economy was in the toilet. A decade earlier, McIntosh had been purchased by a for-profit group that mismanaged its potential.

But the fundamental reason the institution of McIntosh College closed was that not enough members of New Hampshire's public trusted McIntosh to deliver a material benefit. A new generation of Dollys was not interested in applying. When the school's finances frayed, no one on the leadership team was able to parlay its very appealing contract for its students into a sustainable enrollment.

Interestingly, McIntosh was not the oldest New Hampshire institution to be felled by a lack of trust in 2009. It wasn't even the oldest institution to fail within twelve city blocks. Just a mile and a half away, officials at the Saint Charles Borromeo Church, which opened in 1893, were salvaging light fixtures, pews, a church bell, and paintings of the Stations of the Cross for storage, donation, or sale, after the bishop of Manchester announced the church would close in response to a lack of parishioners. Some stained glass would be repurposed for a thirty-nine-unit low-income housing project that would be built on the site, once the church was razed. That year, the bishop signed similar decrees ordering the legal closure of the Saint Denis Church of Harrisville and the Saint Patrick Church of Bennington.

Say what you want about the faults of the church—and I have, and loudly at that—but in many communities, local churches host food pantries, addiction counseling resources, day-care centers, and programs to keep troubled teens off the streets. And nearly 90 percent of the people who use these services are not members of the congregation that offers them. One study found that, in urban areas, each historic church building that continues to serve a congregation generates an average of $1.7 million in economic activity each year.

The 2009 institutional closures did not stop at New Hampshire's colleges and churches. Health providers noted that year that financially strained hospitals had closed five labor and delivery units across the state.

And the Ocean Bank of Portsmouth, which opened in 1854, was closing after 155 years of service.

New Hampshire's woes were a particularly dramatic example of national trends. In 2009, national college enrollment dipped for the first time in decades, from 70.1 percent to 68.1 percent. That year, 375 colleges and campuses, ten rural hospitals, and *thousands* of churches permanently closed their doors, all for the same fundamental reason: they lacked cromulent public support. The average weekly service of a church declined to 105 attendees in 2009, down from 137 in 2000 (a rate of loss that, if continued, would see the last American churchgoer attend their last service in 2048).

In addition to all the tangible goods they provide, colleges, churches, hospitals, and even banks also draw on centuries of hard-won expertise to provide the public with answers to life's biggest questions of science, philosophy, and faith.

In New Hampshire in 2009, as these once-bright cornerstones winked out of existence, they were replaced by business plazas, development projects, or vacancies. As the traditional threads of New Hampshire's communities were withdrawn, people were incentivized to turn to other sources of information about life's big questions. If not through institutions, how were they delving into the mysteries of health, of the universe, and of what, if anything, lies on the far side of death?

Well, there was Andy.

Five

BONES AND SUGAR

Well, where you see one of them blue lights flickering around, Tom, you can bet there's a ghost mighty close behind it. It stands to reason. Becuz you know that they don't anybody but ghosts use 'em.
—Mark Twain, *The Adventures of Tom Sawyer*

Between their protocols and their growing pile of ghost-hunting equipment, the Kitt Research Initiative was beginning to build an infrastructure for itself, from the ground up. Next on the list was a vehicle, something that could transport an investigation team and all their gear.

The best possible vehicle for a ghost hunter is, clearly, a 1959 Cadillac Miller-Meteor Sentinel limo-style endloader combination car with an ambulance conversion makeover. This became fact in 1984, when the vehicle was featured in the movie *Ghostbusters*.

Absent a tricked-out antique Cadillac, the best modern vehicle for ghost hunting is the Chevrolet Tahoe Midnight Edition SUV, marketed to ghost hunters as having "enough room to stack at least eight bodies and a shovel," and a "six-speed automatic transmission . . . as smooth as the satin lining of a coffin." But the Tahoe comes in at more than $65,000.

Andy didn't have $65,000. He was still selling kayaks out of his renovated turkey coop.

"What you have to remember is, I was friggin' broke," he said.

He wound up with a Ford van, covered in white-colored paint and rust-colored rust. The rear seats had been taken out so that the back could function as an equipment cage; once, to power their electronics, Andy installed a mammoth half-ton generator that caused the whole van to tilt heavily to one side.

Over 2009 and 2010, Mike, Beau, and Andy, sometimes joined by an interested friend or two, packed into this van to go on dozens of investigations, usually on weekend nights.

One night in February 2009, the team found themselves at a drab olive-green home in Gilmanton. It was only fifteen years old, but it had been built in the traditional New England style, with second-floor gables and other nods to an architecture of yesteryear.

The family there, relatives of a friend of Beau, had experienced a series of dramatic, if somewhat trite, signs of a haunting—shadows seen from the corner of the eye, tapping noises, and objects popping up where they hadn't been left.

While there, Beau sensed an intrusive spirit and "crossed it," sending it on to a better place.

But before the spirit left, Beau received a clear image of a body buried in the woods behind the house.

"I knew," she said, "there were bones out there."

This was the first time she had received such a vision. When she told Andy, he was excited. He was still learning the ropes of the scientific process, but it was clear that KRI had a hypothesis, with both an independent variable (Beau's vision) and a dependent variable (the chance of there being a body in the ground). It was only a single data point, but the odds of finding a random corpse were so small that it would be compelling proof. With the hypothesis in place, Andy wanted to perform the experiment: digging up those woods. If they successfully used Beau's gifts to find a body, the media would have a field day, and KRI's place in the paranormal pantheon would be assured.

Only one thing lay between them and fame: three feet of frozen February soil.

And so, they bided their time. When summer approached, Andy asked Beau to draw a map of the body's location, noting all trees of at least nine inches in diameter, within ten yards.

That weekend, Andy held the map as he and Mike trailed behind Beau. She walked through the woods behind the Gilmanton house. The trees had leafed out into an unremarkable, scrubby forest. Everything looked more or less the same.

When Beau finally stopped, Andy checked the map. His pulse quickened. The forest Beau had diagrammed matched their setting almost perfectly. Just a single tree was missing.

Andy next took the map to four other places in the forest, to see whether there was something about the growth pattern of trees that would make the map seem accurate everywhere. But none of the other sites matched even remotely.

It was time. They established a ten-foot-by-ten-foot search area.

There were two shovels. Beau sat on the ground, and a new trainee, a young woman who loved horror movies, sat beside her. It was dry and in the mid-sixties, but the labor was intense. Soon, Andy and Mike were sweating heavily.

"Dig!" Beau barked, teasing. "More!"

Beau found it easy to joke around with Mike. The two had bonded over their shared experience in the pub that night. It had propelled Beau right past the zone of disbelief, and into a space of uncritical sympathy for his trauma.

Digging reminded Mike of one of his many unpleasant alien experiences. As a teen, he'd woken up one night and looked out the window to find five aliens digging holes in his backyard. They were three feet tall, with pug faces and big black eyes, dressed in rustic burlap that made Mike think of garden gnomes. The holes they dug were perfect cores, like the sort of hole you might see in a cartoon. After watching for a few minutes, Mike went back to bed.

He believes now that the gnome-aliens were not an accurate representation of the reality outside of his window, but a sort of screen projection, created to prevent him from seeing or remembering what was really happening. During other encounters, they'd appeared much different—mostly five feet tall and gray, as is typical of movie representations, and one that was six feet tall, pitch black with cat eyes.

Once, Mike's wife urged him to seek out mental health services to address his trauma-related stress. Mike tried to tell the doctor about the way he felt without mentioning the aliens. The doctor prescribed an anti-anxiety drug, but it made his stomach hurt and didn't seem to help. And the alien encounters kept happening—like when he lost time while standing on the roadside with friends, or when a King Kong–sized alien peered at him through a second-floor window (he thinks this was another projection). At one point, he said, his day-to-day mental state got slightly easier to live with when an alien tinkered with his brain in a way that allowed him to accept what was happening. In some ways, it was an improvement, but it also felt like a massive violation.

He said the idea that there are good and bad aliens was "bullshit."

"This thing, this plague, has scarred me for life," he told one interviewer. "People can fuck off with all that."

Digging in the woods with Beau and Andy was a bright spot of fun. The fact that they were beginning to take on quests in the service of bizarre beliefs helped to normalize Mike's experiences. Though the fun did lose its luster over time.

"We dug," said Andy. "And dug. And dug."

Time passed. The one hundred square feet began to feel like it was the size of a football stadium. Finally, there was the glorious sound of a shovel hitting something that was neither dirt nor rock nor tree root. The group huddled around their find. It was an old Pepsi bottle, an antique. A little while later, they turned up a milk bottle of, Andy noted portentously, "indeterminate age."

But no body.

They kept digging. Looking at all the unturned earth in their search site, Andy's enthusiasm began to flag. His smoking breaks became more frequent, and longer.

Finally, Andy called it quits. The mapping that Beau had done was, he concluded in a report of the experiment, "powerful stuff."

Though, of course, he wrote, "It would have been far more powerful had we uncovered a body."

Beau said that, though her vision was accurate, ghosts don't experience time in a linear way, so the body could have been there in prehistoric North America, in which case the bones would have long ago moldered away into nothingness. Or perhaps the body would be there in the distant future.

Andy was not one to be put off by a single failed experiment. If Solid Phantoms were easy to document, it would have been done long ago. And you know what they say—if God closes a door, well then, everyone freaks out and blames ghosts.

There were other experiments to conduct. And that was why, one night, Andy found himself in a dark parking lot, flinging small handfuls of a white powder into the air and then quickly picking up his camera to take pictures of the fallout. He was after tiny and ephemeral objects that had roiled the ghost-hunting community for years: orbs.

Orbs are ghostly balls that are rarely seen by the naked eye, but which appear in photos and videos. Some orbs are big. Some are small. Some are stationary. Some float or zoom through the air. They can be almost any color, and some appear to have the features of a face within their sphere. Believers say orbs are visible spirits. Skeptics say orbs are camera flashes reflecting off dust.

Beau was in the spirit orb camp, because it seemed to her that the orbs showing up in KRI photos and videos often correlated with her sense of otherworldly presences.

Andy wanted to explore the possibility, and his sense of the scientific process was getting more finely honed, because he had enrolled in psychology classes in the bachelor's program at Granite State College, part of the same state-funded education system that included the University of New Hampshire. He hoped to transfer to UNH and take up advanced studies in psychophysics, a field that analyzes the relationship between physical reality and the perception of reality once it's been filtered through a flawed piece of hardware: humans.

You can see this in action by holding up two fingers.

"When you look at my fingers out there you see two fingers," said Andy. "You move it closer, you get that weird double-image thing. That's called diplopia."

There are other, less obvious flaws to our perception. For one thing, there is a slight time lag that occurs as the neurons fire to communicate an image to the brain. For another, our eyes fail to pick up things that are obviously there, like gamma rays or magnetic fields.

Influenced by these classes, Andy began to think of orbs as a psychophysics problem. The perception was glowing balls of light, but how was that perception being impacted by the artificial hardware of the camera, and the biological hardware of the eye? He decided to document as many orbs as he could, looking for those attributes that might distinguish the spiritual from the organic debris that is well-known to aficionados of palynology (the study of dust).

Andy's behavior around dust—of both the naturally occurring and Andy-made kind—began to resemble a sort of minor mania. When a piece of light-tinted lint floated through the air over the dark carpet of his apartment, he dropped what he was doing, grabbed a Canon PowerShot S110, and took pictures of it from various angles, turning the flash on and off.

He quickly found that creating orbs was as easy as pie. Definitively easier, in fact, because all it took to make spectacular orbs was a single particulate ingredient of pie, such as confectioner's sugar.

He took pictures of motes of cornstarch, of flour, of finely ground parsley, of cinnamon, of droplets of water from his spray mister. He used digital cameras and film cameras, with and without flashes in a variety of settings and lighting conditions.

Some of what he captured was decidedly weird. One "orb" looked like a big letter P falling from the ceiling. "But it still, it clearly had a gravity-fed trajectory," he told me. "So I got no reason to say that's a dead ghost or a dead person, you know?"

He also came to realize that he could recreate any orb attribute. The color was often simply the color of the material he used—cinnamon was brown, flour white. Flinging irregular motes of parsley created orbs of strange shapes and sizes that proved to be a battalion of Rorschach blots.

The ghostly faces?

"Take a mister with water, spray it on the camera," he said. "You can see faces in 'em. Because you can see a face pretty much anywhere if you really want to."

If a spirit orb is identical in every way to a piece of dust, Andy concluded, there is no evidence that orbs are spirits at all.

Andy and Beau reviewed some orbs that they had captured during investigations, and checked how they correlated to Beau's visions of spirits. Beau still saw a connection but admitted (Andy later wrote) "the correlation wasn't nearly as good as she had hoped."

Even relatively informal, unpublished experiments like this put KRI miles ahead of most groups, and Andy was feeling his oats.

"Belief-based claims are meaningless to empirical science," he wrote in a blog post, "so any group that makes claims it can't support . . . will (rightly) [be] considered irrelevant by those of us who can. . . . So, if your group clings to the claim that you have an orb photo even after you've heard the rebuttal (and can't prove otherwise), our group will consider yours unscientific and irrelevant."

Andy wrote a confrontational and challenging missive to other ghost-hunting groups in the area. If they failed to take the steps that KRI was taking, he said, they were not serious groups.

Andy's academic program required him to take classes in research methodology and to read books, or (better yet) real, peer-reviewed research journals.

"Are you taking classes in research methodology?" he asked other ghost-hunting groups. "Are you reading books, or (better yet) real, peer-reviewed research journals?"

This was how Andy tended to communicate. He was perfectly comfortable shitting upon the willy-nilly practices employed by the vast majority of ghost-hunting groups.

"I am not ParaCentric," he declared. "I am ScienceCentric."

With these sorts of statements, Andy was charting out a tone and culture for both KRI and himself. The group would have high standards and would not suffer fools lightly.

But still. Beau's takeaway was just the slightest bit different. Andy could recreate any spirit orb with dust, that was obvious. But of course, this didn't flat-out disprove them. What about the orbs that correlated with spirits she had seen intuitively? Did the strict science prism that Andy envisioned for KRI leave any room for her personal experiences?

These questions, about the relative weight of facts and personal experiences, would prove to be important. For Beau, for Andy, for Mike. For KRI. And for all of us.

Six

SPACE BETWEEN LIVES

> *Despite the fact that the memories she had cherished were now become hideous things, she sought to drag them forth and compare them, ruthlessly, with what must have been the treasures.*
> —Winston Churchill, *The Dwelling-Place of Light*

Beau was always happy to barter her services to advance KRI's mission, or to help instruct others in the way of mediumship, as she had taken to doing through a series of intuitive development classes.

Her standard fee for clearing a house of a restive spirit was several hundred dollars, but she sometimes did it for free, if the homeowner allowed KRI to first come in and investigate. Once, she answered a phone call from a local paralegal named Antje Bourdages.

The previous night, Antje had heard a metal clanking sound coming from downstairs, and found that her keys had mysteriously fallen from a dish in the center of the countertop to the floor. In the morning, Antje's young teenage son entered her upstairs bedroom, visibly shaken, and led her to the kitchen. A set of Tupperware was laid out on the floor in a neat row, like a parade of colored turtles swimming across the linoleum. No one in the family knew how they got there.

It was an odd little enigma. Beau asked if Antje would, rather than pay the fee, play the role of psychic guinea pig for her students. Soon, Antje stood in front of a class of Beau's students, aspiring mediums who practiced divining information about her case. Some were clearly off base, but others seemed on point.

"I see a child involved," said one.

"A child has been scared," said another.

"I hear a clanking metal sound," said a third.

There was no definitive explanation for the Tupperware-shifting spirit, but Beau followed through on her promise to clear the space. Sometimes this was like scrubbing a small blot from a piece of clothing, but in cases like this, she subjected the space to the equivalent of a heavy-duty wash cycle—she cleansed Antje's entire house of unwanted psychic stains. It worked. Antje never had another unexplained incident in her home.

In June 2010, Beau addressed a ghost in the home of Ed Lane, a Somersworth hypnotist, as part of a barter deal. In exchange for clearing his home, Lane would use hypnosis to help Beau, Andy, and Mike access memories of their past lives. Beau thought of "good old Ed" as a "cool dude."

Andy said Lane was a nerd. A dork. In fact, "a nerd dork." But not in a bad way.

"He's overly cheerful and excited about stuff," Andy said. "I don't know how he does it, but he does. There's nothing about him that's bad."

Lane planned to bring Andy to memories from a past life, and then fast-forward like a movie, up to the point of death. And then, using a regression technique he'd only recently learned, beyond. The space between lives—presumably the space inhabited by spirits—is an incorporeal state of existence shrouded in mystery. Not even hypnosis could access it reliably, and Andy sensed that Ed was just as excited to peer into its depths as Andy was. A woman Andy knew had been taken back to a previous life as a Prohibition-era gangster who died in a shootout, but couldn't be brought beyond the point of death. Andy was eager to try.

Once he sat in Lane's chair, he went down easily enough. So easily, in fact, that when Andy listened to the recording of Lane hypnotizing him, Lane hypnotized him again.

"I get hypnotized every time I listen," Andy later said.

Lane began with a bit of unrelated business. He had helped hundreds of clients quit smoking or other harmful behaviors. He told Andy that every time Andy saw the color red, he would be more determined to quit smoking.

With that suggestion percolating, Lane took Andy to a past life—working on the deck of a wooden ship. He was a sailor.

"What's your name?" asked Lane.

"Bill."

Those who work in past-life regressions are often on the lookout for details that might lead to a positive identification of a historical person.

"What's your last name?" Lane asked Andy.

"I don't know."

When Lane asked Andy to write down his last name. Andy scrawled an X on the sheet of paper. Lane, stymied, moved on.

With prompting, Andy remembered a maritime storm springing up. Andy was high above the deck, working on the rigging, when he was knocked loose by a gust of wind. "My shoulder got caught in the ropes as I went down, which, uh, shredded my shoulder," Andy said. "Everything just got pulled apart." The ship docked in England, and Andy remembered lying on his side in terrible pain.

"I died in bed," Andy said.

Now came the moment that Lane was really after. With a renewed sense of intensity, he urged Andy forward in time. When Andy crossed the threshold of his own death, he had an experience that he has struggled to describe ever since. It was a moment of immense knowledge, and connection. "Reality as we perceive it is literally a consensus of agreement," he later said. It also served as another confirmation of a feature of the afterlife that his father's spirit had signed off on—there, all time is experienced simultaneously.

"What, exactly, do you feel?" asked Lane, trying to delve into the particulars of that space between lives.

"It's kind of like coming home," Andy responded.

He didn't expand on this to Lane, but Andy didn't mean coming home in a spiritual sense. He meant that exiting a human body into the hereafter was like getting home from a long day of work and plunking down into an easy chair, exhausted. Living sucked.

Lane seemed intensely curious.

"How do you manifest the life you want?" he asked. Andy took it to mean that Lane was asking what he, Lane, could do to manifest a different life for himself—er, Laneself.

"He really, really wants to know how to manifest a better life," Andy later told me. "What he doesn't wanna realize is that he already has, but it ain't beginning for a while."

Andy broke the bad news to Lane.

"Dude," said Andy. "Your powers of manifestation end at the moment you're born. You get to pick everything you want about your life, then you're born. And it all goes to plan."

Andy said he could hear the frustration in Lane's voice.

"What do you see?" Lane asked. But Andy's experience wasn't about seeing. It was about being. And the disconnect between the question and his current reality jarred him out of his rapture. Their session was over.

Afterward, Andy went outside, looked at something red, and lit up a cigarette. He was colorblind.

He knew, from Lane, that it often takes multiple sessions for anti-addiction hypnosis to work.

But as he puffed away, Andy realized that there was only one thing worse than an unsuccessful smoking cessation program. And that was a successful smoking cessation program.

"I kind of enjoy smoking," he said. "So I never went back."

When it was Mike's turn, Lane took a different approach. He suggested that, instead of a past life, Mike would benefit most by accessing that lost memory from when he was four years old.

"I wasn't going in there to relive that at that moment, ya know?" Mike later said. "But I was open to it."

For the first time in more than thirty years, Mike had a chance to access the full memory of his childhood encounter with a UFO.

Andy looked on with interest as Lane put Mike under hypnosis and walked him through the evening of that 1983 family gathering in South Hampton. In his mind's eye, he once again watched a flying saucer with a belt of otherworldly lights rise above the house, casting it in a red backlight. The craft moved to within a few yards of Mike and his cousin, and then zipped upward as quickly as if it had been slurped up a straw by a giant. Ed guided Mike into the seam in his memory, that instant in which he seemed to have blipped into a different position.

Mike began to describe newly surfaced memories. He was ascending something like an escalator into a doorway in the craft.

"I think I'm in it," said Mike.

"You're in the craft now?" asked Lane. His voice sounded more high-pitched and reedy, contrasted with Mike's deep gravel.

"I . . . I feel like there's things around but I don't . . . I don't see anybody," Mike said.

"When you say things you mean beings," clarified Ed.

"Yeah," said Mike.

A minute later in the conversation, Mike caught a glimpse of a humanoid.

"I didn't really see his face. He moved past me and he was almost hunched when he walked. He . . . he . . . wherever it is, that wouldn't have been his normal thing, I don't think. It . . . there was a physical attribute, to be hunched. I don't think it would be comfortable to be like that all the time," said Mike. There was a tension in his voice now.

"Okay. Was the form human like? Arms, legs, torso, head?" asked Lane.

"Yeah . . . legs, I didn't catch. I'm assuming he must have had them but it was more like hips up. The arm, if you looked at a human arm it seemed like the wrist was longer from like where your wrist should be. Instead of going to your hand it was almost like it was just another piece of forearm there. It could have just been the way he held it. It looked almost crunched when he went through."

A few minutes later, Mike reported that he was lying down, though whether on a table or floating in the air, he couldn't say. There were beings around his feet. He had a sense that they were celebrating, and that he had been here before. Sitting in Ed Lane's chair, Mike began to grow visibly upset. On an audio recording of the session, I could hear him sniffling.

"Again, do you get any sense of them communicating with you on any level? Telepathically? Orally—er, auditorily?"

Mike didn't. But his sense was that he was being subjected to a checkup. Not a medical checkup, but some sort of assessment of change over time. Andy could see tears were now rolling down Mike's cheeks.

With Lane asking more questions, Mike seemed to realize that the activity around his feet didn't have to do with him at all.

"Any idea who it's for?" Lane asked.

"My cousin," said Mike.

"It's your cousin, okay," said Lane. "And what's happening with her?"

But that was when Mike's emotions overpowered Lane's ability to control the session. Beau said Mike was out of his mind with fright.

"When you heard him, like, scream in panic, it was kind of bone chilling, man," she said. "It was really uncomfortable to hear."

Beau said that she felt the pain even more deeply because she wasn't scrutinizing every word to determine whether Mike's account was real. "I believe in every core of my being. I believe him, and that made it harder to hear."

Mike was no longer in a hypnotic state. The upsetting experience jarred him out of his trance. He sat in the chair, sobbing. His session was over. And he never underwent hypnosis again.

"Whatever it was was severe enough that he didn't want to look at it still," said Andy. ". . . Whatever it was snapped him out, and he could never go back."

Seven

POWERFUL COSMIC FLUIDS

> *The barbers, all over the world, are a shrewd, observing race; their occupation brings them into close contact with the surfaces most sensitive to the mesmeric influence; and they are, therefore, very likely to have become possessed of the secret of Mesmerism at an early period, and perhaps it has descended to them as a mystery of their craft.*
> —James Esdaile, *Mesmerism in India, and Its Practical Application in Surgery and Medicine*

If Mike or Andy or Beau were interested in whether hypnosis-induced memories were admissible in a court of law, they could have made a short drive to Concord to ask any number of legal professionals employed at the Franklin Pierce Law Center, New Hampshire's first and only law school.

There, they might have learned that, legally speaking, the idea of using hypnotic regression to access lost memories was on extremely shaky ground. Back in the 1970s, federal courts cautiously allowed it as admissible evidence, but in the late 1980s, a series of cases in which regression therapy led to unsubstantiated accusations of childhood abuse led the courts to rethink their position, allowing such evidence only on a "case by case" basis.

Had they wanted to ask Franklin Pierce faculty about this topic in early 2010, they would have felt a sense of urgency.

That's because, just like McIntosh College, Franklin Pierce was in an existential crisis that further jarred New Hampshire's higher-education landscape. Notwithstanding a respectable body of derisive lawyer jokes (What's the proper weight for an attorney? Three pounds, not counting the urn), having a flow of credentialed lawyers in New Hampshire was an acknowledged public good.

Like many law schools, Franklin Pierce helped the community by linking the talents of its law students to the needs of the broader public; in this case, Franklin provided free services that helped usher investors through the patent process, protected them from predatory and fraudulent patent services, and connected them with resources that could help them bring their inventions to market. The school also maintained the only academic library in the nation dedicated to intellectual property law.

This facilitation of economic growth was a kindness to the market, but the market was not being kind to Franklin—or any of the hundreds of other law schools that provide pro bono representation for everything from falsely accused inmates on death row to Native American communities suffering socially unjust levels of toxic pollutants in their environment.

Despite these public goods, law institutions were losing support across the nation—enrollment had peaked at 147,500 students nationwide, and was beginning a decline that would see a 25 percent reduction, to 110,000 JD students, over the next several years.

Public institutions are supposed to be a bulwark against such losses—and in the case of Franklin Pierce, this promise was actualized when the University of New Hampshire stepped in to absorb the law school. Though Franklin Pierce didn't publicly disclose its finances at the time, within a few years the public would soon learn that UNH operated it at a deficit that would grow from $1.2 million to $6.7 million annually over several years.

But of course, UNH isn't tasked with profiteering. No public university is. On behalf of the taxpaying public, they have the freedom to advance missions that go beyond making money. They are an investment from the public, mediated by elected representatives, based on the understanding that universities enable upward social mobility for millions of Americans. And that's a good thing. When it lasts.

In the wake of their hypnosis sessions, Mike, Andy, and Beau could have also learned more about hypnotism from UNH, which has actually had a hand in shaping our modern scientific understanding of what is, at its core, a really bizarre practice.

UNH's psychology department has taught thousands of students about hypnosis, largely due to Professor Ronald E. Shor, who joined UNH's psychology department in 1967. Shor was an international authority on hypnosis and cognitive processes, and published more than one hundred books and articles.

He helped untangle the modern history of hypnotism, which began in the 1800s, when an up-and-coming philosophy of thought—namely, science—was trying to sort all sorts of medieval beliefs into buckets of truth and myth.

Giant lizards flying through the air? Truth! (French naturalist Georges Cuvier documented the pterodactyl in 1800.) People with perfectly perpendicular foreheads tending to be stupid? Myth! (The pseudoscience of phrenology, which assigned mental characteristics to corresponding bumps on the skull, was blown up in 1815 by French physiognomist Marie Jean Pierre Flourens.) An elemental difference between things that live, and things that don't? Myth! (Friedrich Wöhler debunked the belief of "vitalism" when he synthesized urea in 1828, proving that people and animals are made up of the same elements as rocks and air.)

And on and on and on.

Hypnotism, on the other hand, was a cagier beast to categorize.

As Shor and other academics have it, the seeds of modern hypnotism were planted when a German physician named Franz Mesmer began entrancing patients in Vienna as a means to relieve mental conditions. This became known as *mesmerism*.

Practically nothing Mesmer thought about hypnosis was correct. He thought that he was directing his own powerful "cosmic fluids" into the subject's mind. He thought that a magnetic force was vital to the process (causing him to coin the term *animal magnetism*) and used a big piece of metal, such as an iron bathtub, to facilitate his sessions. Mesmer aimed to violently push his patient's mental state into a period of crisis, which often led to a period of calm that Mesmer would claim as a win.

Mesmer's story is a case study in the incestuousness of Europe's social elites—seemingly everyone you've ever heard of from that era just constantly chummed around with one another. In this case, Benjamin Franklin (yes, that Benjamin Franklin) invented a musical instrument in which fingers rubbed thirty-seven rotating colored glass bowls on a horizontal spit, producing the ethereal chiming sound we today associate with wine glasses at parties that are not too stuffy, yet certainly not too fun. Franklin called the instrument an *armonica*, a term that worked very well (for everyone except, of course, London East Enders).

Mesmer, of Vienna, then working as a doctor, used Franklin's instrument as a musical backdrop for group hypnosis sessions, during which (mostly female) patients gripped onto the iron tub while handsome male assistants clasped their legs together from behind and applied pressure to their chests, which culminated in the women having a series of "convulsive fits."

Mesmer became so proficient at the armonica that when he hosted a social visit from sixteen-year-old Wolfgang Amadeus Mozart (yes, that Mozart), the musician began incorporating the sound into his musical works. Mesmer's treatments were taken up by Marie Antoinette (yes, that Marie Antoinette) and then drew the ire of King Louis XVI (yes, that Louis, or possibly not). The king established a royal commission to investigate Mesmer's claims, and the commission was chaired by Benjamin Franklin (yes, that Benjamin Franklin, again). It also included Dr. Joseph-Ignace Guillotin (who created, yes, that guillotine, much to the chagrin of the aforementioned Antoinette and Louis).

The commission undertook its work by lying to patients about whether the supposedly critical magnets were present. One woman fainted after drinking water that she thought had been magnetized (but wasn't) and was revived by water she thought had not been magnetized (but was). Another woman went into fits when she was told (falsely) that a practitioner was directing magnetic force at her through a door.

When Franklin's commission declared Mesmer's ideas to be erroneous, Mesmer was cast from society and lived out of the public eye until his death. Today, the whole incident is remembered as the creation of the first placebo-controlled blind trial (yes, that placebo and that blind trial), which has underpinned so much of science ever since.

This is the official record of hypnotism's history, but Andy's understanding is very different. He told me that Mesmer's powers were, up until the point of King Louis's intervention, demonstrable and dramatic.

"He was getting astounding results," said Andy. "I mean, people developing psychokinetic powers and, you know, all sorts of just amazing abilities."

Institutions, Andy said, were the cause of Mesmer's downfall. The general threat of Mesmer's teachings to the power structure of his day resulted in a discrediting campaign.

"Mesmerism was pretty much abandoned," said Andy. "The ESP effects were dropped from the hypnosis process. And modern hypnosis developed out of the core of what survived."

This does not comport with the institutional account of things. Which definitively proves whichever of the two viewpoints you prefer.

The end of Mesmer was the beginning of a fresh inquiry into hypnotism. The commission had debunked the element of magnetism. That cleared the way for a new understanding of hypnosis, one that would be shaped by science.

Post-Mesmer efforts to harness hypnotism for the public good continued. In the 1840s, Scottish surgeon James Esdaile performed 1,300 surgical operations in India using only hypnotism as an anesthesia. After entrancing patients, Esdaile dropped hot coals on their legs to ensure they were without pain. He then went ahead with arm, breast, and penis amputations, dental surgery, scrotal surgery, and tumor removals. In one memorable instance he claimed to have painlessly removed a tumor "7 feet in circumference and 2 feet at the neck." Needless to say, had Esdaile tried withholding anesthetic from his fellow English oppressors rather than from the oppressed people of India, he would have likely been tarred and feathered on the very first day.

At UNH, Shor wrote that hypnosis subjects were thought to be capable of all sorts of mystical mental feats: clairvoyance, mediumship, foreknowledge of the future, and medical diagnosis.

But one of the most frustrating things about hypnotism is that it works better for believers than skeptics.

"Thus investigators with cooler heads were often at a disadvantage," Shor wrote, "whereas those with wilder fantasies and blinder resolve were more readily able to produce striking effects. There was consequently a natural selection toward extravagance."

And that extravagance was often embodied by stage performers trying to one-up each other. For example, hypnotists sometimes instructed their women subjects to enter a lion's cage, at least until 1890, when one of those women was fatally mauled by one of those lions during a carnival in southern France.

But behind the freak show, scientists continued to nibble away at the actual causes and uses of hypnotism, seeking consensus. One by one, a chain of researchers conducted experiments and wrote papers that advanced the practice from cosmic fluid to a harnessing of the hidden contours of the mind. Today, hypnosis has gained wide acceptance in scientific and medical circles. In a clinical setting, hypnosis can alleviate pain and anxiety, and is thought to be a good complementary treatment for weight loss (though there is less evidence that it helps with smoking and other addictions).

But we know to be suspicious of hypnosis-surfaced memories thanks in part to people like Shor, who found that hypnotized UNH students recalled events no more accurately than unhypnotized students. He concluded that "posthypnotic amnesia is not a case of state-dependent retention, nor does hypnosis provide retrieval cues that can lead to the emergence of previously unrecalled memories."

But what does that mean for people like Ed Lane?

Hypnotism is a great example of a practice that has developed along parallel paths. Inside academic institutions, an ever-more-cohesive, evidence-based framework guides practitioners, who are also held to a set of professional and ethical standards.

But outside academic institutions is a rough-and-tumble free-for-all, in which knowledge is accrued by experience and personal connections, and ethical practices are loosely guided by a combination of the market and the criminal justice system. The individual practitioner, rather than the collective, is the biggest factor in what claims are made, and what ethics are adhered to.

Ed Lane received his training at the Thomas Institute of Hypnosis, which was founded in Manchester in 1997 by Cynthia Kile Thomas. The school is unaccredited, and at the time Lane attended, it gave four-day and seven-day courses for fees in the $700 range, in which students could retake all classes and tests until such time as they earned their certificate. Thomas in turn got her certification from an organization in St. Louis, Missouri, founded by Don Mottin. And Don Mottin's résumé in the field was essentially that he was hypnotized one time while serving as a Marine in Japan in 1972. After he became a police officer, Mottin taught himself to hypnotize people, and used the techniques to do exactly what science suggests cannot be done: help witnesses recall details about crime scenes.

And so, Mike was regressed by someone who had been trained by someone who had been trained by someone who taught themselves hypnosis in an effort to make hypnosis do something it's not designed to do.

But perhaps that's not a fair assessment. There was a broader question in play in New Hampshire, where institutions were giving way to a more fractured landscape of self-appointed experts. Was it possible that this was actually setting the stage for a triumphant age of citizen science, led by people like Andy, Beau, and Mike?

Tragically, Shor is no longer around to weigh in on this question, or on things like between-life hypnosis sessions. On a Friday in late January 1982, at fifty-one, his body was discovered around 7:30 a.m. by a university police officer at Hersey House, a psychology research center on the UNH campus. It was a reported suicide.

After a career spent in pursuit of objective science, Shor had become just another piece of grist for cheap thrills for UNH student ghost hunters, who believe that his spirit roams Hersey House still.

Eight

MORE BITE

The empirical scientist, however, looks at the 30 heads in a row and truthfully states that there is probably something wrong with the premise.
—Andy Kitt, Ghosts, Science, and the Number 42

In 1989, Valerie Lofaso, thirteen, was walking down the hallway between classes at her junior high school in Derry, New Hampshire, when a girl she barely knew hit her from behind with a flying karate kick, sending Val to the ground with the wind knocked out of her. The worst of it wasn't the kick. The worst of it was the way the crowd of kids standing all around did nothing to help.

Val's lowly social status in 1989 made sense. She had entered elementary school with the determinedly positive daffiness of a Luna Lovegood.

"Weird and open and loving. That's who I was," Val said. "I was always joyful and singing and dancing on the playground. At all times."

This didn't go over well with the other kids. As her social situation fermented, she found new friends. Exciting friends. She crept through the stench-filled living rooms of pandemic victims in Stephen King's *The Stand*, and raged at the Children of Darkness who kidnapped the true love of the vampire Lestat. When Darcy said that Elizabeth was not handsome enough to dance with, she bristled. When Christopher Pike's deadly chain letter began circulating around a group of teens, she puzzled over the culprit's identity.

This dynamic—childhood tormentors pushing, books pulling—nudged her by degrees into a world of fantasy. Even then, she had what she thought of as an unusually strong, and sometimes wearying, intuition about the people around her.

"If someone can see through a facade, that makes people who put up a facade very uncomfortable," said Val. As a teen in Derry, she pursued thrill-seeking paranormal activities that were typical of the day. She played with Ouija boards. She lit candles and chanted verses. She was occasionally light as a feather, stiff as a board. For her peers, it was a goof. For her, it was a serious search for knowledge.

The karate kick in the hallway was just one more thump in a constant drumbeat of emotional and verbal abuse. What sustained her were dreams of being a writer herself.

Between classes, she wrote fiction, and the things she wrote were often rooted in the books she enjoyed most as a reader. Ghosts. Vampires. Period pieces centered on dramatic interpersonal relationships.

Val married in her mid-twenties. Her husband brought her to a house in a gated community in southern Florida to join a partner in a start-up marketing firm. At first, it seemed like a fresh start. But it proved to be isolating, and her husband's attentions went from tender to cold.

Almost immediately after the birth of their daughter, Juliana, her husband checked out emotionally. They separated five months later. After the divorce, Val felt ill-equipped for her unexpected role as a single mother. "I felt guilty every time I had to buy a cheap pair of jeans at the store. Food was a concern," she said.

Plagued with bad thoughts and mired in a potentially fatal state of depression, there was, at times, just one thing keeping her alive. Every Wednesday night at nine, she put her unpaid utility bills aside, told herself she wasn't hungry, and switched on the Sci-Fi Channel to watch ghost hunters Wilson and Hawes approach potential hauntings with a sense of logic and reason. This was an immensely powerful experience for her. For her entire life, she had been hiding her true interests, because she was afraid people would laugh at her. But *Ghost Hunters* was an expression of paranormal interests that was more socially acceptable than laughable. She credits the show with literally saving her life. One of the best things, she said, was that she wasn't the only person watching.

"I knew," she said. "I was no longer alone."

She and Juliana moved to an apartment back in Portsmouth. She found part-time work at a local fish market. She was receiving unemployment and intermittent rent assistance from the state.

On weekends, Juliana's school district would send her home with bags of oatmeal and pasta and tomato sauce and other staples that would get them through the weekend. Val hated her job, she couldn't pay her bills, and she was terrified that her deficiencies were hurting her daughter.

All along, Val kept writing, writing, writing. She finished a draft of her ghost-themed teen novel, and began to shop it around. Doors weren't slammed in her face so much as they stood mutely, resisting her efforts to initiate contact with the anonymous publishers and editors behind them. When Val did finally get a response from an editor, it was a rejection. But he deigned to give her feedback, and that in itself seemed like a minor miracle. She was told that her draft needed "more bite."

More bite. Her mind worried at the phrase, trying to translate it into actionable items that would transform her manuscript into something marketable. *More bite.*

In order to describe paranormal events, perhaps she should better acquaint herself with the world of spirits. That realization led her to SPRG, just as it was being imploded by infighting.

Andy and Beau wanted a different future for KRI. Along with Mike and Andy's girlfriend, there were only four actual members, which allowed for a more intimate and cohesive unit to uphold protocols. Val was one of many supporters who tagged along on investigations, but only as a guest, with no voice in how to conduct the business of spirit-hunting. Almost immediately after Val began attending, Andy and his girlfriend split up, and KRI needed a new member. Andy and Beau were trying to decide between bringing on Val or another candidate, named Marisa. Andy quizzed them.

If you flip a fair coin ten times in a row, and it comes up heads all ten times, what are the odds that it will come up heads on the eleventh coin flip?

The ability of this question to elicit irrational responses has led the science community to brand it the Gambler's Fallacy—evoking images of bleary-eyed addicts pumping a continuous stream of money into a slot machine because it is "due" to pay off.

Fifty percent, said Marisa. She held a science degree.

Val had never heard of the Gambler's Fallacy.

Ten times? In a row? she thought. *Something isn't right there.*

"Probably it will come up heads again," she said.

And that was the answer Andy was looking for. He wanted someone who would look at a series of incredible coincidences and see, not random statistical noise, but something to be interrogated. He wanted someone who was willing to question the premise.

When Andy began talking to Val as a possible replacement for his ex on the KRI team, he was also doing reconnaissance on the possibility of a replacement for his ex in a more intimate sense. When Andy and his girlfriend broke up, Beau told him there was a new girlfriend waiting for him in the wings. Was that Val? He wasn't sure.

Andy and Val went to dinner at Coat of Arms, to talk about KRI, and her status as an investigator.

Val's years as a social outcast had molded her face into a carefully arranged state of repose: her mouth was always hinting at the slightest twist of amusement, her first line of social defense. Andy thought she was beautiful.

From Val's side of the table, Andy's amorous intentions were obvious, though whether her intuition on that score was paranormal in nature is debatable. A romantic connection was not something she was entertaining.

After the dinner, Val had a private conversation with Beau. She wanted to be an investigator, and she was open to a friendship, she told Beau. But nothing more.

"Would that screw up my chances to be an investigator?" Val asked Beau.

Absolutely not, Beau told Val.

Beau knew that Andy lacked the knack for putting women at ease. Once, after a successful event, he'd smacked Beau's ass, which led to an exhaustive debate on the differences between congratulatory jubilation and exploratory libidinization.

Andy soon understood that in Val he had a fine new investigator, and an empty bed. She was the right choice, he said afterward.

And Val found that Beau's assurances bore out.

"I never got any pressure from him," Val said.

Andy said that he was struck by Val's buttoned-upness.

"It was brutal how self-contained she was," he said. He speculated that this was an expression of her being insecure.

Val didn't think of it as insecurity.

"I've always had a lot of self-confidence," she said. "It was other people I was unsure of."

But perhaps, at long last, KRI would prove to be a softer landing spot for her.

With Val on the team, they continued to investigate reports of spirits—at graveyards, in homes, in bars, and in the woods. Each investigation was a sort of a murder mystery, but instead of trying to figure out whodunit, they were often trying to figure out whoitdun.

By the summer of 2010, Val was just one of many pieces that were falling into place for KRI. They had the equipment, the van, the training, and, increasingly, the experience and critical thinking skills that eluded so many aspiring ghost-hunting groups.

Andy was consistently impressed with Beau's talent. She had visions of bolts of fabric in a building that turned out to have once been a dressmaker's shop. She revived a catatonic young woman by clearing her of an oppressive spirit. As news of her talent spread, Beau was getting inundated with requests for her living-room readings, from those whose hearts beat quickly at the thought of communing with their dearly departed, and also from those whose hearts had ceased to beat altogether. It was becoming a bit of a whirlwind.

"I can't do this at home anymore," she realized, a couple of weeks after Ed Lane's hypnotism sessions. "People are knocking on my door, um, inappropriately."

Andy saw this as an opportunity for the whole KRI team. What if they brought all of their talent and expertise and equipment and contacts within the community into a single, physical location? What if they had a brick-and-mortar home that could serve as a launch point for all that they wanted to do?

It could be the nucleus of a bigger dream—the Kitt Research Initiative as a school, with Beau the spiritual leader and Andy the administrative head. Val and Mike could be administrators, or faculty. It would, in a time of declining institutional trust, be a new kind of institution, one that was relevant for the future.

They began looking for a rental space that could serve as a place for psychic readings, science experiments, equipment storage, lectures, and classes in the paranormal.

The search eventually led Andy to consider a lease at Gowen's Corner Plaza in Greenland, New Hampshire, right across the street from the Winnicut River. The Winnicut is a New England quirk; it runs just nine miles and flows north, a cranky Yankee defiance against the general tilt of the North American continent.

The plaza already hosted a hodgepodge of businesses—a bike shop, a dance studio, a start-up gym, and Diamond King Sports Cards, which sold collectibles.

Andy walked through a self-consciously neutral commercial unit. The front hallway led to a cluster of offices, a larger oatmeal-on-macaroni bland conference room, and a modest bathroom.

The whole space was missing ceiling tiles, but that didn't bother Andy in the least. He had acquired a deep reservoir of workman's skills, and wouldn't mind doing some renovations. One of the rear offices, he noted, was capacious, large enough to hold a bed. The lease prohibited renters from actually living here, but surely they wouldn't mind him taking an occasional nap here, or even, in rare instances, sleeping through the night. Or, let's be honest, living here.

For Andy, life in the turkey coop had been very uncomfortable lately. The landlord there had become consistently surly with Andy. Something about the landlord's dog, which had suffered a very traumatic incident, in tragic circumstances. While Andy was reversing down the driveway in one of his clunkers. There were veterinary bills at issue.

Andy signed the lease. For $800, KRI had a home. And so did he.

Nine

VERY BAD BEDFELLOWS

For there's my wife, now, can't smell, not if she'd the strongest o' cheese under her nose. I never see'd a ghost myself; but then I says to myself, "Very like I haven't got the smell for 'em."

—George Eliot (Mary Ann Evans),
Silas Marner: The Weaver of Raveloe

In late 2010, shortly after Andy signed the lease for the KRI offices, Beau got an email from a woman named Marilyn McGehee, who had found Beau's name in an online directory of professional psychic mediums. The email challenged Beau to have her mediumship tested in a more formal scientific setting.

McGehee was working as a research assistant on a research project by Karla Baker Curtis, a PhD candidate at Holos University Graduate Seminary in Missouri.

Holos is not, I should mention, highly regarded in the academic world. With a focus on spirituality, it was founded and operated by Norm Shealy, a Duke University–trained neurosurgeon who got interested in energy healing and went on to sell a series of New Age products like "Crystal Bliss Oil" and a $1,600 "Pulsed Electromagnetic Frequency

Machine" that purportedly does everything from regenerating damaged tissue to combating depression.

The fact that the call for Beau came from a fringe entity like Holos, rather than a research institution like UNH, is, on its surface, odd. After all, shouldn't America's universities be devoting resources to explore the paranormal phenomena that are of fervent interest to millions of Americans? It's difficult for academics to argue that the issue is not worthy of study, given that they have undertaken such frivolous questions as whether the Midwest is actually flatter than a pancake (Texas State University and Arizona State University joined forces in 2003 to prove that Kansas is indeed flatter than a flapjack).

One reason that mainstream research institutions aren't (for the most part) taking up studies of the paranormal could be found within the psychology department at the University of New Hampshire, the very department from which Andy hoped to learn psychophysics. The faculty was dominated by people who took a dim view of the paranormal: Associate Professor Bill Stine, the resident psychophysics expert, who naturally assumed that visions of ghosts could be explained by the human mind; the newest chief lecturer, Mark Henn, who had cut his teeth on a dissertation that showed self-confident students were most likely to think that they could exert telekinetic powers on the roll of a die; and Henn's advisor, Professor Victor Benassi, who helped to mold incoming instructors in his role as head of the UNH teaching center. Benassi is one of many who have wondered about the relationship between science and the paranormal.

"How should the scientific community respond to the upsurge of popular interest in the occult?" asked Benassi, in his research. "The interest in occultism and pseudoscience may in fact be dangerously distracting from rational solutions to social problems and corrosive to rationality."

Over the years, science and the paranormal community have proven to be very bad bedfellows. Institutional scientists have undertaken countless studies that attempt to link a belief in ghosts to mental deficits or pathologies. Some researchers have tied poltergeist reports to brain damage (of the part of the brain that manages visual processing), while experiences with "shadow people" or "being watched" are associated with certain forms of epilepsy.

But it's not clear that paranormal believers, writ large, are in any way pathological. A 2022 review of more than seventy published research papers by one team found that while paranormal beliefs coincided with an "intuitive thinking style," they weren't convincingly linked to any cognitive deficits.

If anything, the sheer numbers of believers suggest that paranormal beliefs are a feature, not a bug, of the human animal. In 2023, Pew found that a majority of Americans have been visited by dead loved ones. Humans naturally believe in all sorts of different things for which there is no objective evidence—that a demonstrably harmless spider the size of a pistachio represents a mortal threat, that modern violent crime rates are high when they are in fact historically low, or that the orange candy circus peanuts taste good.

Science-driven research universities are not the only institutions that treat paranormal beliefs as a punching bag. The church has long been shaming mediums for violating biblical tenets not to speak to spirits, and many leaders within the church take a dim view of ghost hunting. Media institutions, too, have been lampooning believers since at least the mid-1800s, when a London newspaper, the *Saturday Review*, went on at length about the outfits worn by apparitions.

"How do you account for the ghosts' clothes—are they ghosts, too?" it asked scornfully. "What an idea, indeed! All the socks that never came home from the wash, all the boots and shoes which we left behind us worn out at watering-places, all the old hats which we gave to crossing-sweepers. . . . What a notion of heaven—an illimitable old clothes-shop."

More recently, in the 1970s, an emerging community of skeptics began debunking and ridiculing paranormal claims, driven in part by the notoriety of celebrity stage performers like James "The Amazing Randi" Randi, and Penn Jillette (of Penn and Teller).

With institutions calling them crazy, heretical, and silly, it's not surprising that paranormal believers have evolved a culture that places little stock in those systems. Faced with an ongoing attack on their belief systems, people like Andy have harsh words for the skeptic community in particular, which he says has become so biased as to categorically reject any evidence that suggests paranormal activity.

"James Randi is a professional liar. He's an arrogant little midget," said Andy. "He's got little big man syndrome." (Randi was 5'6". That is also Andy's height.)

Andy also said that Penn Jillette, of Penn and Teller, is "notoriously skeptic by volume. He's right because he's loud."

Andy also noted Penn's height.

"He's like six feet," Andy said. "He's a big dude." (Penn is actually 6'6".)

People with scientific backgrounds who delve into the paranormal with good intentions are shocked when they get sucked into internecine warfare. Elizabeth Mayer, associate clinical professor of psychology at the University of California, Berkeley, began studying the paranormal in 1991, when a rare, hand-carved harp stolen from her daughter was recovered by a man who used a dowsing rod. Mayer was used to academic disagreements, but she found the level of discourse in skeptic-themed publications to be atrocious.

"Reading the *Skeptical Inquirer* was like reading a fundamentalist religious tract," she wrote in a book on the topic. "I found the journal dismayingly snide, regularly punctuated by sarcasm, self-congratulation, and nastiness, all parading as reverence for true science."

Does what amounts to an academic disagreement over the efficacy of dowsing have to be so ugly? Maybe it does. Mayer says skeptics and believers are locked into a zero-sum game in which the fundamental truth of one view strips all legitimacy from the other.

"It shakes foundations," she wrote. "I think it may be as basic as any fear there is. . . . In the face of fear like that, no wonder rational consideration of apparently anomalous experience is so elusive. No wonder the scientific establishment looks the other way, moves elsewhere as fast as it can."

And so, believers and the scientific establishment are like doomed lovers; believers crave the reassuring embrace of lab-tested affirmation, but any time a scientist dives into an experimental bed with a psychic, it doesn't end well. The spurned believers decamp to exotic beaches to drink mason jars of watermelon vodka and coconut rum, and complain about how science just doesn't get them. Science, for its part, scrolls through endless spiritual Instagram posts, wondering why no one is asking it to parties and whether it will die alone, and whether that is, indeed, the end.

Very Bad Bedfellows

And that's why Beau was being contacted by Holos, a college at the extreme shallow end of the academic credential pool, rather than Harvard. Curtis and her colleagues had designed a triple-blinded, small-scale experiment that was meant to be scientifically airtight.

Beau agreed to participate immediately. Perhaps the outcome would be the Solid Phantom.

Beau was staying in a hotel when the call came. "This is for Michelle," Curtis said, identifying the deceased. "What do you pick up on this?"

Beau had only that first name, communicated by Curtis. Curtis couldn't accidentally give Beau leading hints, because Curtis also didn't know anything about Michelle. The clients were recruited and questioned by McGehee.

A moment after Curtis spoke, Michelle's spirit appeared in the hotel room with Beau. She looked to be about thirty-four, a well-dressed brunette with a slight curl to her shoulder-length hair and some blotches or light freckles on her face that she covered with makeup.

Beau and five other mediums, all working independently and from different locations, would each do four readings for bereaved clients. Each client would choose between the reading that was intended for them and a reading done by the same medium for another client.

If clients chose the "correct" reading more often, it would demonstrate that the readings had intrinsic value.

After describing Michelle to Curtis, Beau asked Michelle other predetermined questions for the study. *How would you describe your personality? What were your hobbies or favorite activities?*

The answers that Beau related back were rich with detail and character. Michelle was a smoker, and a gossip who liked to go out rather than stay in the house. She had a connection to Harley motorcycles and danced, either tap or ballroom. She was verbally abused by her husband, but didn't leave. "I just took it," Michelle's spirit said.

Beau asked the last question.

What was the cause of your death?

Cancer, Michelle told Beau. At the end, Michelle was in a hospital, with very blue veins, with some sort of problem in the left side of her chest. Either her breast or her lung, Beau wasn't sure.

"There were some issues with an IV," Beau told Curtis. "Something went wrong with fluids that went into her body via IV, either a missed injection or an infection."

Michelle showed Beau a scene of her deathbed.

"It was nighttime, dark, and very quiet in the hospital," Beau said. "She showed her sister sitting in a chair at the end of the bed, with her feet propped up on the bed."

After the readings, another research assistant took transcripts back to the clients for evaluation, creating another level of blindness. The clients would also give a rating of how accurate the reading was.

When the client—who was indeed one of Michelle's two living sisters—received Beau's reading, she gave a glowing response. Michelle was thirty-eight years old and died of cancer, having undergone the hair loss and the strikingly blue veins that Beau described.

The problem with the IV was "EXACTLY accurate," wrote the sister in response. "This hospital scene is EXACTLY accurate."

But overall, the outcome of the Curtis study was disappointing. While slightly more bereaved clients chose the "correct" medium reading, it was not a statistically significant difference.

However, Beau's individual results were more impressive.

All four of Beau's clients in the study chose the correct reading, and Michelle's was rated "mostly accurate" by her sisters. Despite these direct hits, other things Beau said did not seem to match up. Michelle's spirit had said she was verbally abused, but took it. Her living sisters saw it as a relationship of equals, with mutual bickering. Other details, like the dancing and the motorcycle, were also not accurate.

And Beau's other three readings were rated lower by the clients, indicating that they were chosen simply because each was the best of two less-than-ideal options.

A scientist reviewing the study would conclude that Beau was simply the medium who was lucky on that day. But a believer might ask a different question. Were the other mediums in the study actual mediums? Beau had been recruited simply because she self-identified as a medium. Without any communal standards or certification process for the craft, there is no way to confirm that any of them were legitimate.

This was the inherently tricky nature of KRI's quest for a Solid Phantom. If no one can point to a distinguishing feature of an orb, how can one study legitimate orbs? If no one can point to a distinguishing feature of a medium, how can one study legitimate mediums?

There's another complicating factor, too. Academic reviews show that researchers who believe in mediumship are more likely to find evidence than nonbelievers. This of course means that those who believe are biasing their own research, unless this of course means that those who *don't* believe are biasing their own research.

Finding fundamental, and replicable, constants of the spirit world—this was the sort of problem that KRI wanted to solve. After Beau was tested by Curtis, Andy began talking about doing another version of the experiment, one with larger sample sizes. He began to look for volunteers, but soon put it on the back burner in favor of a more pressing matter.

Andy, Beau, Mike, and Val were about to open the doors of the Kitt Research Initiative Center for Consciousness Studies to the public for the first time.

Ten

WHAT IS THE SOUL?

Down dropt the breeze, the Sails dropt down,
  Twas sad as sad could be
And we did speak only to break
  The silence of the Sea.
    —Samuel Taylor Coleridge, *The Rime of the Ancyent Marinere*

To pay the rent and other operating expenses at the KRI Center, Andy and Beau mapped out three revenue streams. They would sublet offices (including one to Beau to work in, and one to Andy to sleep in). They would create a retail space to sell books, crystals, and other spiritual items. And they would hold regular events to generate ticket sales and donations.

One of the early conditions Andy set was that there had to be a standard of truth for those who presented information at the center.

"We host events ranging from serious paranormalia to spiritism so fringe-like you have to see it to believe it," he wrote on the KRI website. "But we host nothing we won't back up either empirically or through rigorous personal validation."

They sent an invitation to roughly 250 contacts, advertising an open house on the first Saturday of May 2011. At the appointed hour, dozens of vehicles began parking on the cracked concrete, disgorging mediums of all shapes and sizes—grizzled men who felt certain they were guided by wolf spirits, women who felt a deep kinship with angelic presences, engineering geeks who wanted to prove ghosts existed, engineering geeks who wanted to prove ghosts did not exist, goth twenty-somethings who wanted to know whether vampires were real, and wrinkled seniors who could read tarot cards as easily as the blue plate menu in the local diner. Familiar faces in the crowd included the hypnotist Ed Lane, and Antje Bourdages, she of the Tupperware-moving spirit—both among those now filing into the main hallway.

Inside, they saw how thoroughly Andy had transformed the space; he'd installed drywall, electrical wiring, and ceiling tiles. In one spot, to improve the traffic flow, he'd knocked out a doorway-sized hole in an office wall, adding struts to help support the weight, which he could claim as a victory for as long as the wall stood (otherwise, he said, "the apartment upstairs would be in our main room, you know?").

Not all of the work was for visitors; that Andy had moved into the Center was particularly apparent in the tiny bathroom, where he had custom-drilled and installed a kitchen countertop that included a squirt hose and an extension panel that could be dropped over the toilet to hold plates and bowls.

"I gotta be able to wash my dishes," he explained.

There was also a curtain blocking off roughly half of the four-by-six room, behind which was a wiring panel, and a large plastic tub. Every day, Andy would squeeze the tub into the narrow space between the sink and the wall, and stand in it to wash himself, using a shower attachment he'd installed. It was arduous, but Andy was game. He was on a mission.

The main meeting space was Andy's masterpiece. Anticipating lectures and classes, he'd completely rewired the room to support an overhead projector and additional banks of lights controlled by a new panel of switches. He'd even installed dimmable accent bulbs into floor-level heating vents, which gave it the feel of a theater.

The once-nondescript front wall was now sky blue, with a fringe of grass painted at the bottom. The other walls he painted tan, and then he used some of his mother Betty's old stockings to make a darker patchwork of brown around the bottom edge, representing the texture of soil.

What Is the Soul?

"When you're sitting on the floor, it's like you're sitting in a field," he said.

The response ranged from overwhelmed to underwhelmed. When Antje Bourdages described the brown swirls that dominated the room, she said, "It looked like somebody had explosive diarrhea on the walls."

A few minutes before 1 p.m., the crowd sat, expectant, on rows of folding chairs turned toward the front of the room. Andy and Beau had decided that each KRI member would give a lecture.

Val made her way awkwardly to the front of the room, wondering why she had agreed to speak first. And why she had chosen a topic that she, honestly, knew nothing about.

The crowd quieted as she began.

"What is the soul?" she asked.

It was the subject of her presentation, and it was also a question that, up until recently, she had no special knowledge of. She'd googled it as frantically as a seventh grader cramming CliffsNotes.

She was meant to speak for forty-five minutes, followed by a fifteen-minute Q&A. She paused portentously.

"Where is it located?" she asked her audience. One minute down.

Val had come to KRI to support her novel about teens and ghosts, but it had quickly evolved into a journey of self-discovery. She was taking Beau's latest twelve-week intuitive development course, and had been drawn to Beau's charisma. "She had a way of, just everybody liked her," Val said. "You couldn't help but like her."

But Val's admiration was, for reasons she did not understand, not fully reciprocated.

"Val liked Beau a lot more than Beau liked Val," said Andy.

And now, standing at the front of an expectant crowd, an internal storm cloud was shading Val's inner dialogue. Beau was the expert. She should be speaking, Val thought.

I don't deserve to be here. What do I know? Why would anybody want to listen to me?

In stilted speech, Val awkwardly walked the audience through an experiment conducted in 1907 by a physician in Haverhill, Massachusetts, named Duncan MacDougall. He rolled the beds of dying nursing-home residents onto a large scale, hoping to demonstrate that at the moment of death, when the soul presumably departed the body, it would register a change that would translate a metaphysical concept into physical terms.

Because MacDougall reported that one of the six patients he measured lost 21 grams at the time of death (an effect that other doctors immediately chalked up to a loss of liquid associated with a tendency to sweat and overheat at the time of death), an idea took root that 21 grams is the weight of the soul.

As Val spoke, the audience was silent. If her presentation was a Christmas tree, she wouldn't know which side to hide in the corner. She fumbled her way through a series of other concepts of soul from cultures around the world. The best thing about the presentation, she later said, was that it came to an end.

"I don't know if I flubbed it, or just, I was speaking just to speak," she later said. "I wasn't speaking on anything that meant anything to me."

Her duty done, Val slunk away from the lectern and tried to lose herself among the snacks at the back of the room, where Mike's desk held a sheet cake with the new KRI logo on it. There was no admission fee, but they had thought to make a marked donation box available—it had dense fiberboard walls, with a magnetic lid that could be flipped open at the end of the night to count the haul.

People were congratulatory, though Val suspected they were merely taking pity on her. One of the people who struck up a conversation with her was Antje, who offered a kind smile and praised her on her presentation. Val was polite, but cautious. Her junior and high school experiences had left her wary of friendship—particularly with women.

"I have a long history of being bullied and abused by females in my life, from my mother to people that I thought were friends. It's been a struggle to me, to trust women," said Val.

But the vibe here at KRI was reassuring. Weren't they all, in some way, outcasts by virtue of their beliefs? As Antje spoke, she exhibited a friendliness and openness that Val found hard to resist. Soon, she found

What Is the Soul?

her defenses lowering, almost of their own accord. If Val had flubbed a lecture, but made a friend—well, that wasn't such a bad trade-off, was it?

Val and Antje watched as Mike stood before the room, presenting on alien abduction. Some quiet people avoid speaking because they are bad at it. But Mike was surprisingly articulate. Something about his blue-collar workaday vocabulary, and the deliberateness of every word he spoke, carried its own charisma.

A Center wall held a corkboard calendar full of events, most with a ten-dollar suggested donation for entry. Mike would host two twice-monthly events that reflected his interests—alternating Fridays at seven, it was "Social Saucers," which was a safe space for people to share their experiences and theories, or to ask questions. On the odd Fridays, he would host "Crypto Connections," an exploration of Bigfoot and other cryptozoological creatures. There was, at KRI, an affinity for alliterative event titles that bordered on the fetishistic. Other events that would come and go included Andy's Anomalies, the Lifting Lodge, Galactic Gazing, Martian Matinee, Mike and the Medium, Sacred Space, and the Psychic Sampler.

Val's name was conspicuously absent from the list of event hosts, but she made up for it by working on some of the organizational aspects—she organized a Didgeridoo Journey Circle (hosted by Joseph Carringer, a professional didgeridoo musician and sound therapist), as well as a Sunday healing clinic with various spiritual healers. In this way, Val was finding a role where she could help fill the calendar by supporting, rather than presenting.

"After that opening day," she said, "I pretty much avoided doing lectures."

As the inaugural open house kept going, some people left, but more arrived, leaving the big meeting room with the fancy lighting and soil-brown or diarrhea-brown walls near capacity at all times. Beau followed Mike's discussion with a presentation about how death is emotionally processed in Western culture. For this, she drew on not only her readings of books, but her readings of spirits.

Many of the attendees had come specifically to see Beau, who had experienced a spike in popularity due to the recent publication of a book she had authored. It had begun when, while writing a memoir-style draft of her early spiritual experiences, she learned something about herself.

When it comes to writing books, Beau said, "I am a math major."

So she asked Andy, who had a journalism degree, to edit it. He read what she had written, and said it was bad.

"Not even close," he told her.

So she rewrote it. He read her second draft.

"You do know the word okay is spelled with an 'A' and a 'Y,'" he said.

"Okay," said Beau. She had not known that. "Cool."

She rewrote it again. And he edited it again. After months of back and forth, in January 2011, Beau had finally released *Cracking Open: Adventures of a Reluctant Medium*. It was published by Runestone, a tiny company owned by a local medium Beau knew. The book laid out Beau's origin story—how she'd been a skeptic until her grandmother's spirit appeared and introduced her to spirit guides, who had nudged her deeper and deeper into developing her intuition. And how, finally, her guides and her grandmother urged her to embrace the responsibility of helping spirits to move on from their earthbound twilight existences.

Beau was gaining more followers than ever before, but she continued to chafe at the tensions between serving her living clients, who paid the rent, and her dead ones, who truly needed the help. For a while, she only offered readings via email; one question from the client, one emailed response from her, and she moved on.

The promotional material for the book included a blurb from Mike declaring it a must-read, while Andy contributed a foreword.

"Here I am, attending the University of New Hampshire to learn sufficient experimental methodology to test Beau as completely and scientifically as is humanly possible," he wrote. He added that, when it comes to paranormal phenomena, "Beau lives them every day while I tag along trying to describe and define them."

What Is the Soul?

The only KRI member whose name didn't appear in Beau's book was Val. Val, who also had written a book. Val, for whom authorship was a lifelong dream. Val, who had been sitting on her draft for years. She watched from the sidelines with a keen interest.

The always garrulous Andy, whose every conversation amounted to a lecture on something-or-other, had saved the final open-house speaking slot for his own presentation, entitled "What Is Paranormal, Anyhow?" He wanted to provide a foundational education on the differences between psychic phenomena, the paranormal, metaphysics, and parapsychology.

As I would come to learn through conversations, this was classic Andy fare. As he educated himself on the paranormal, he was struck by the lack of agreed-upon categories and definitions for the most fundamental concepts. In other areas, he found maddening inconsistencies.

For example, many in the community shared Andy's belief about spirits experiencing all time simultaneously. But few seemed to understand the ramifications of that belief.

"Well, that means that we don't live our lives to learn anything," Andy once told me. Which directly contradicts the understanding of reincarnation as articulated in Hinduism, Sikhism, Buddhism, and Jainism, all of which describe a soul that revisits Earth to attain perfection.

"People hated me for this for years, and some still do," Andy said. "It's really, uh, essentially, we live lives to do the whole Disneyland thing. You know, we want to experience something new and that's what we're doing."

This fatalist view even applied, he told me, to our current conversation.

"Our interaction right now was predetermined," Andy told me. "There's nothing we can do to change it, you know? I mean, I can pull out a gun and shoot you, and if we're supposed to talk in an hour, you will not die. You know?"

Andy laughed. "You might bleed all over my carpet," he continued, "but, you know—"

I interrupted him.

"Please don't test your theory," I said.

"I don't think my gun actually works," Andy replied.

Absolutely not the response I was looking for.

By the time the KRI open house was over, the members glowed with enthusiasm. There had been a large crowd. There had been applause. The donation box had cash in it.

But over the coming months, Andy would learn that the revenue streams were not as robust as he had hoped, at least not right away. To help make it through those first few months, he sold off his meager stock of five kayaks, with one fetching a price of $2,000.

Despite the shaky finances, the establishment of the Center acted as a glue for the KRI members. No longer were they meeting up for investigations and then heading off in different directions. Now there was a place to just hang out and bullshit endlessly about the afterlife—and about life in general.

Their shared interests were drawing them into a close comradeship. Andy and Beau were always talking about how to push KRI into the future. Val began a hot and heavy friendship with Antje, who had started to show up at some Thursday night discussion groups, and enrolled in Beau's intuitive development course. Val and Antje would meet outside of the Center and have long talks on the phone about the nature of spirits, their own journeys under Beau's guidance, and their lives. Antje seemed to understand the struggles of motherhood and daughterhood, the sadness, the pain. She offered Val a consistent sympathetic ear about Val's troubles—her poverty, her mothering challenges, her frustrated writing aspirations. And Antje admired Val's openness, her intelligence, and her authenticity. When Antje sold her house, Val helped her move.

"She was calling me her soul sister," said Val.

And everyone loved Mike, the one guy who always seemed to notice who needed a kind word, or a helping hand, and never asked for anything in return.

Val was also getting closer to Andy. Mindful of the big role Andy's editing had played in Beau's book, Val asked Andy to take a look at her draft of the first entry in what she envisioned as a series: *Tangled Web of Friends*. Her protagonist, Josie, was a teen theater geek who finds herself on an extended camping trip beset with adolescent angst and dangerous ghosts. He found it to be a much different editing experience than he'd had with Beau. Andy saw in it real writing talent. "Oh my God," he said. "She's the best instinctive writer for her genre that I've ever seen. You know, the editing is fixing commas."

This was uplifting for Val—the friendship with Antje, and getting thoughtful, positive feedback on her book. On Andy's birthday, she showed up with a big round chocolate cake that said "Happy Andy Day" in clumsy blue icing, ringed by marshmallows and Hershey's Kisses. Whenever team members parted ways, they began saying goodbye with a light "love you."

Collectively, they decided that their next big priority was to open the retail space. Andy said that, in addition to dowsing rods and space-cleansing sage candles, it could sell Beau's book, Val's book—which they would publish—and a book that Andy was working on to explain some of the technical ins and outs of ghost-hunting equipment, plus some of his broader thoughts about the nature of evidence and science in the field.

There were no suitable retail spaces in the Center, but a stroke of luck came when the frame-maker next door moved out. The big, open space there was perfect for a retail operation and only cost a few hundred dollars a month. Andy decided to rent that as well. All it needed to become the perfect retail space was a bit of renovation, and Andy was happy to once again put his workman skills to good use for the cause.

Eleven

THE SPIRITUALIST WIND

Do avoid "spirits." I cannot bear to think of you as engaged in a course of wickedness and deception. Indeed, Maggie, it is very sad. Say so to Kate.

—Dr. Elisha Kane, letter to his betrothed,
Margaret Fox, circa 1853

When the old white Ford van gave up the ghost, Andy's postmortem suggested that it was murdered by the half-ton gas-powered generator that provided electricity to the team during investigations. A shock absorber in the chassis had exploded, thrusting a leaf spring into the heart of the wheel well, which proved to be fatal.

"This vehicle is done," said Andy. "I need something else."

He could put together a thousand dollars, and not a penny more, to replace the van. When he asked Beau to give him a ride to a cluster of used-car dealerships along Route 1, she agreed, though she was busier than ever. As the spring of 2011 ripened into a sweltering summer, institutions seemed helpless to quiet a restive planet. There was a nuclear disaster in Japan, floods along the Mississippi River, violent uprisings against governments in Greece, Spain, Libya, Syria, and Egypt, and a cholera outbreak in Haiti. The country of Belgium seemed unlikely to survive a populist political movement that demanded breaking the country up into a series of smaller city-states. A lot of people were asking Beau whether the world was about to end. The basic idea sprung from the Mayan Long Count calendar, which on December 21, 2012, would shift from one set of *baktun* cycles to the next. Given that each *baktun* comprised 394 years, the end of the thirteenth and final one in a set was a big event in Mesoamerican timekeeping.

We know that the idea of an imminent Mayan-predicted doomsday is nonsense thanks in part to UNH archaeologist William Saturno, who in 2010 uncovered the room of a Mayan record keeper that included references to dates seven thousand years in our future.

Still, doomsayers suggested that, at the appointed hour, Earth would be swallowed by a flaring sun, or collide with an imaginary planet, or get sucked into a supermassive black hole at the center of the Milky Way.

All these theories were rejected by astronomers, but no one in the paranormal community was calling astronomers. Nor were they calling William Saturno. They were calling Beau. Thankfully, her spirit guides had been reassuring about the planet's short-term prospects.

"This idea of the world ending, it's like, it happens annually," Beau said, practically rolling her eyes. The world was not ending, but shifting, she told people. A spiritual shift, away from the doctrine of organized religion and toward a more enlightened era.

One of the people she told this to was Val, who remained a bit in awe of Beau, of her spiritual talent, and of her matter-of-fact steadiness in the turmoil of people pressing her for answers.

What Val didn't fully understand was that Beau's apparent unflappability was, at least in part, an emotional mask, crafted since childhood to make people like her. Her mother had instilled in her a mantra of denial.

"You never say what's wrong. You never complain. You don't ask for help," Beau said. In this, she was the polar opposite of Val, who was obviously, and perpetually, stressed by her struggles as a single mother beset by poverty, and as an aspiring writer seeking her break.

"If you asked her how her day went, you know, you could almost imagine without hearing her open her mouth, saying it was miserable," said Andy.

The Spiritualist Wind

Val's mother had taught Val to speak in the language of complaints. If she went to the wedding of a cousin, she would come home with a story that included only the bad things—the hitch in the proceedings, the rapidly declining attractiveness of an aunt, the worst food item in the buffet.

Sometimes, it seemed like Val labored under a perpetual dark cloud.

Beau protected herself from problems with a smile that minimized the space for pain, trying to wade to a better psychic space. Val's protection strategy was to exhume each detail of each problem, hoping to expel them from her psyche. Neither approach was perfect, and they were also mostly incompatible.

"That's where we didn't click," said Beau. "I think I was a little envious."

But Val was similarly envious of Beau, the financially stable epicenter of KRI's spiritual inquiries whose feathers remained unruffled by every downpour of bad news.

This was why Beau hadn't warmed to Val. She later told me that she liked Val. She just didn't like interacting with Val.

"If she ever needed anything from me, I'd be there for her," said Beau. "If she wanted to go out for coffee, I would say, 'Nah.'"

When the van died, Beau and Andy drove to Hampton, where a whole ecosystem of car dealerships sopped up revenues from the heavy traffic flow. They were reputable and disreputable, offering new and used vehicles, with a whole spectrum of complex financing options. Anybody could get a past-its-prime car here, but it took an expert to get a good deal.

Andy had that expertise. Many of his childhood memories took place in his father's auto shop in Syracuse, the one with the purple tow trucks. Andy hoped to snag a hidden gem among the beaters due to be auctioned for $300, the standard minimum.

Dude, Andy would say. *I'll give you four and you don't even have to take it to auction.*

As they got into neighborhoods dominated by dealerships, Andy told Beau to turn left.

"No," Beau said lightly. "Why don't we just turn right?"

"Whatever," Andy remembered saying. "You're driving. We'll go right before we go left and you can see what's down there."

A little ways down the road, Beau said, "Why don't we check this place?"

Andy was dismissive. It was an Audi dealer, the lot dominated by new vehicles, the upper end of the market.

"This is definitely not what I'm talking about. There's nothing here that I'm interested in. There's no freaking way."

"No," Beau agreed, "you're right, you're right." But she was parking while she said it. She led Andy to a dealership across the street. It was small, with several dozen cars on the lot, and had just the air of seediness that Andy wanted. Andy had once worked at a Mazda dealership, and so the old blue Protege sitting near the back of the lot caught his eye. It was a manual, which Andy preferred, as his father had before him.

"Why would you get an automatic when you know you can buy a stick and it's half the money? Why would you do that to yourself?" Peter Kitt would growl. "What are you gonna do, you spin out of control in an automatic? You wind up in a ditch. If you spin out of control in a stick, you can downshift and come right out of it."

Andy called a middle-aged car dealer over to the Mazda. Not for sale, the dealer said. It had a blown transmission.

"It's a stick," said Andy. "You can't blow the transmission."

The dealer invited Andy to sit behind the wheel. It started fine, but as soon as he put it into drive with the clutch off, it died.

That's not a transmission problem, Andy thought. *It kind of sounds like a vacuum leak.*

They agreed on $800. Andy managed to drive the car home by feathering the gas and beating on the clutch pedal. The following day, he used twenty cents' worth of vinyl-rubber hose to fix the seal, and it ran slick as slime. The vehicle was perfect for him.

At first, Beau didn't tell Andy that she'd guided him to the Mazda by taking directions from an unfamiliar spirit. Later, when Beau saw the same spirit again, she asked him who he was. He tore open his shirt to show her a zipper on his chest.

"It's my mark of manhood," the spirit told her. "Tell Andy that."

When she conveyed this to Andy, it all fit. Of course his father would use that particular phrase, as Andy's mother had about Andy's college-contracted hickey. Andy took the zipper as a nod to his father's various heart surgeries.

Andy soon learned another detail that made him even more certain of his father's hand in the car search. Andy thought that the Mazda Protege was blue. But he was colorblind; he couldn't see the paint job's red tones. It turned out that the car was a distinctive shade of purple—the same shade as his father's fleet of tow trucks.

Andy's inability to distinguish certain colors was another example of the psychophysics principles he was learning in class. His colorblindness was a slight hardware tweak that altered his perceptions, and therefore his reality. Weren't mediums the same exact thing, people with a hardware tweak that added dead people to their visual spectrum?

Andy considered this, with mounting excitement.

"If they can sit there and establish that the experience of color is not subjective at all, then I can use the same methods to establish that mediums are objective, are observing an objective phenomenon," Andy said. "The actual methods don't exist, but the pattern of methods do exist."

Andy took these sorts of thoughts into his morning classes at Granite State College, where he was continuing to take classes as an adult.

"There's ways of numericizing this stuff, of applying numbers, which is what all of the research is, just finding ways to numericize this, so you could make a decent analysis," Andy said.

He shared his thinking with Bill Stine, the UNH psychophysicist under whom Andy hoped to study. Stine seemed to respect Andy's grasp of the principles, though when it came to Andy's ideas about spirits, Stine all but patted him on the head. He was highly skeptical that applying these methods would yield any proof of the paranormal. The burden was on Andy to gain entry to the program, and then make it happen.

People concerned about the link between institutional distrust and the rise of paranormal beliefs might take comfort from the idea that institutions have evidence-based facts on their side. They think that using these facts to debunk the existence of ghosts is the key to persuasion. This idea was also compelling during the mid- to late-1800s, when churches and academics joined to debunk the heck out of ascendant mystic forces.

That ascent began in 1849. Roughly fifty years after Franz Mesmer popularized the idea of hypnotism, sisters Kate and Maggie Fox, aged eleven and fifteen, told a tale that would spark America's first vibrant spiritualist movement.

At 7 p.m. on a March evening, they stood onstage in the newly built Corinthian Hall in Rochester, New York, where four hundred people had paid a quarter each for the privilege of seeing the two girls. As the crowd watched, the young girls, pale-faced and seemingly nervous, asked questions. Was there a spirit in the vicinity?

Rap!

It was difficult to pin down the source of the sound. The girls on the stage hadn't twitched a muscle. One rap meant yes, they said. Two meant no. Was the spirit afraid to communicate?

Rap! Rap!

These were sounds to upend a generation. The demonstration was so successful that some saw in the performance an end of the universe as described by either church or science. Rochester town officials swiftly appointed an investigatory committee, which inspected the girls before, during, and after subsequent rapping sessions without finding the slightest indication of fraud.

The resultant fame of the Fox sisters sparked a Spiritualist movement that within just five years included roughly 10 percent of all Americans. Thousands of self-identified mediums, mostly disenfranchised women, emerged, many of whom also went on to great prominence and success.

Over the fierce objections of religious leaders who condemned mediumship as a means to commune with the devil, and scientists who said every instance was a con, a whole supernatural economy sprang up, one dictated by free-market enterprise rather than institutional leaders. There was a brisk trade in talking boards, spirit photographs, séances, mesmerists who retooled hypnosis performances into spirit chats, mourning jewelry, and planchettes, among many other products. Adherents coalesced into Spiritualist churches, congregated in semireligious Spiritualist summer camps, and read dozens of Spiritualist publications.

The Spiritualist Wind

Belief was so widespread that it began to color institutional decisions, as in the late 1800s in West Virginia, where testimony from a ghost helped convict Erasmus Shue of murdering his wife, Zona Heaster Shue. Her death was presumed to be natural until her mother claimed that Zona's ghost had visited her on four successive nights and described how she had been murdered by Erasmus. When Zona's body was exhumed, her neck was found broken, and a jury convicted Erasmus of murder.

Every bit of this Spiritualist wind sweeping the nation could be traced back to the Fox sisters, whose ability to elicit raps from ghosts made them America's first nationally celebrated mediums. For forty years, the sisters fanned the flames of the grassroots movement by performing séances and giving public demonstrations of their powers; sometimes, they talked to spirits five times a week.

What was interesting about the Fox sisters was not whether they were real or frauds—what was interesting was that, despite repeated accusations of fakery, they continued to be widely embraced by the public.

A whole host of academics and church leaders demonstrated that the sorts of raps produced by the Foxes could be produced by quotidian means, including the cracking of joints, and using knees and ankles to rap on the floor from behind the cover of long dresses.

Others debunked the Fox sisters specifically.

Salem-born chemist and devout Christian Charles Grafton Page (who would go on to patent the first reliable means to disconnect and reconnect an electrical circuit at the flip of a switch) visited the Foxes twice. In his 1853 account of the encounters, *Psychomancy: Spirit-Rappings and Table-Tippings Exposed*, Page displayed a vocabulary that could send the brightest linguaphile scrambling for a copy of the OED. When I read that "our quondam enthusiast we found at his matins," I paused to learn that *quondam* meant former, and *matins* meant a morning prayer. Within the space of a few pages, I also learned the terms *hebetude* (dullness), *seriatim carneous* (flesh-colored), *instanter* (at once), and *caviled* (made petty objections).

Page, posing as a believer, contrived a series of tests. At his request, the Fox sisters tried to identify the name, cause of death, and city of death for three deceased people. They could not. They tried to produce raps while standing upon a big pillow rounded with cloaks on the floor, and could not. While sitting at a table, the sisters could not produce raps when someone was looking under the table. Page had seen enough.

"This newfangled philosophy," he wrote, "is a poisonous, though covert, fang secretly gnawing at the very root of Christian faith."

And Page was far from the only credible critic. In 1851, a relative of the Foxes revealed that she had been called upon to aid and abet the sisters in their deceptions, by squeezing their hands to indicate knowledge that the Foxes did not know. In 1857, the Fox sisters tried and failed to collect a $500 prize offered by the *Boston Courier*, after the judging committee found that the raps were coming from their feet and joints. During an 1887 investigation, a panel of University of Pennsylvania academics found that the raps were emanating from the Foxes' feet, and one member felt Maggie's leg pulsating with the effort.

As if that wasn't convincing enough to believers, an even more potent debunker emerged: the Fox sisters themselves.

In 1888, forty years after they began communing with spirits, a nervous and pale Maggie took the stage once again, at the New York Academy of Music, with Kate looking on supportively. Maggie, fifty-five, her voice thick with emotion, delivered an unambiguous message to the two thousand in attendance.

"When I began this deception I was too young to know right from wrong," she said, stretching her arms upward. "I hope God Almighty will forgive me and all those who are silly enough to believe in Spiritualism."

Maggie had signaled in advance that she intended to disavow Spiritualism from this stage. But Spiritualists suspected that some sort of skullduggery was afoot. If she had indeed lived a lie for forty years, where was the proof?

Maggie had anticipated the objection. Wearing a black robe, she discarded her shoes, and stepped in stockinged feet up onto a little pine table near the center of the stage. Three doctors emerged from the wings, and knelt around her, sometimes touching her toes.

And then she produced a sound that made the believers in the crowd cringe.

Rap! Rap!

The Spiritualist Wind

It was the sound that had begun the Spiritualist movement forty years ago. And now, the same sound would perhaps end it, coming, as it did, from Maggie's big toe. She cracked the joint repeatedly, so loudly it could be heard in the rear of the auditorium, as the doctors nodded along gravely.

When some in the audience applauded, it seemed to dispel Maggie's nervousness, and she danced about, clapping her hands in relief. "It's a fraud!" she shouted. "Spiritualism is a fraud from beginning to end! It's a trick! There's no truth to it!"

This was a believer's worst nightmare.

"The Spiritualists in the audience almost frothed at the mouth with rage as they left the building and muttered furious threats against their foes," a local newspaper reported.

Maggie gave more particulars in interviews with newspapers and with skeptic writer Reuben Briggs Davenport, who used it as the basis for his book *The Death-Blow to Spiritualism*.

Maggie and Kate revealed that the rapping started as an innocent childhood prank on their mother, but that their mother had become so emotional over the initial demonstration that it was awkward to confess. As the mother brought in others to see Maggie and Kate perform, they became impossibly mired in the lie. And then their sister, Leah, who was twenty years older than Maggie, took them under her wing and turned them into a public act, after which there was a profit motive.

"No one," wrote Davenport, "who does not love illusion for illusion's sake—better, in other words, than he loves the truth—can, after reading this volume, remain a follower of Spiritualism and its hypocritical apostles."

When he wrote that sentence, Davenport thought that the important bit was the forceful debunking. But it was, ironically, the caveat that has gained historical resonance. Because, in the wake of Maggie and Kate's confession, people proved that they did in fact love illusion better than the truth.

After the Fox sisters confessed, the Spiritualist movement grew, not only in America, but in Europe. When Maggie died, four years later, penniless and in the throes of alcoholism, she left behind a world of eight million devoted Spiritualists, more than at any time since her public disavowal.

The Foxes proved that paranormal beliefs don't grow because of the evidence that supports them. And they don't die because of the evidence that contradicts them.

Something else was clearly at play. But what?

Twelve

IN A SUIT OF ARMOR ON ROLLER SKATES

Did Christ live 33 years in each of the millions and millions of worlds that hold their majestic courses above our heads? Or was our small globe the favored one of all?
—Mark Twain, letter to Olivia Landon, 1870

New Hampshire's three-hundred-odd state-approved historical markers are a stone-mounted, copper-stamped paean to institutionalism. Taxpayer-funded markers celebrate the state's first public school, the state's first public high school, and America's first organized ski school. Other markers point to the birthplace of the Seventh-Day Adventist Church; the Worsted Church (known for worsted-wool decorations in the interior decor); and, in Claremont, the oldest standing Episcopal church in the state.

Markers also honor the sites of visits by Presidents Washington and Jackson; the home of a meat-packer named Samuel Wilson, who was the basis for the iconic patriot "Uncle Sam"; and, in Concord, the ratification of the US Constitution.

The idea of adding an alleged UFO abduction—Betty and Barney Hill, in 1961 in Lincoln—to the state's list of historical honorees is, if you think of it, kind of boggling.

And yet, this quixotic quest is what brought Mike Stevens to KRI in the first place. As a young man, he had stolen public signs on a lark; now, he put everything he could into having one erected for posterity, despite a lack of funding, support, or particular knowledge of how markers happened.

The New Hampshire historical highway marker program is jointly administered by the Division of Historical Resources and the Department of Transportation. The criteria are fairly strict; markers must have had "significant impact" and, over time, "demonstrated historical significance."

In addition to a modest petition requirement of twenty NH citizens, applicants must submit written permission from the hosting landowner, a draft of the proposed text, a written historical overview that places the subject in its context, scanned document excerpts from primary and secondary sources, a bibliography, letters of support from experts, ancillary forms, and a written statement from the local town that the marker will not create undue traffic or other safety concerns.

Admittedly, some historical markers are quirky, as in the case of a marker for the house of Winston Churchill (the early twentieth-century novelist, not to be confused with the Winston Churchill people care about). But when Mike's application landed on the desk of the state officials, the alien abduction it described represented a new level of quirk to be considered.

One night in 1961, the Hills were driving through the New Hampshire woods, headed home to Portsmouth from a Niagara Falls vacation. After noticing unusual lights in the night sky, they rounded a curve to find the road blocked by a flying saucer. They were taken aboard, prodded, and probed by short, gray humanoids with no hair, big eyes, and slitted noses.

The Hills were considered to be very credible, because they were civic leaders and professionals in their hometown of Portsmouth, and even more tellingly, because they did not seek out publicity in the immediate aftermath of the incident.

There were also several tantalizingly physical pieces of evidence: circular shiny spots on the back of their car that caused a compass needle to go haywire; a pink powder and rips on Betty's dress; scuffs on the top of Barney's shoes, allegedly caused when he was dragged up the ramp of the spacecraft; and a star map that Betty drew from memory that bore a resemblance to an actual star system about which she had no knowledge.

And yet, there were aspects of the case that also provided rich fodder for skeptics. Barney seemed to be less declarative about the experience than Betty. Much of the abduction was not remembered on the night in question, but surfaced a few days afterward in a series of highly lucid dreams that Betty experienced. And more of the details came out even later, through a series of hypnotic-regression sessions administered by a therapist.

None of the physical evidence led to anything conclusive.

By the time the story broke publicly, in 1965, the car was gone. The pink powder had dissipated from the dress. The fabric was analyzed in a handful of different laboratories to inconclusive effect. The shoe scuffs were just that—scuffs, without a verifiable cause.

Among the skeptics of the Hills' account is Andy, who said the case has more smoke than fire.

"There's a lot of evidence, but it's all trivial," said Andy. "It's like, yeah, the trunk of the car is magnetized. I will not tell you this, but the trunk of *your* car is probably magnetized, you know, from the magnetic field generated at the time of its manufacture."

Less well-known is the saga of the Hills in the years after their fifteen minutes of fame faded.

Betty came to believe that she could send mental messages to the aliens, and encourage them to pilot their craft to a specific location. A network of legitimate scientists and UFO enthusiasts formed around the Hills. They spent several nights at Betty's family farm in Kingston, to see if aliens that Betty had invited would show up. They never did.

One day in 1969, Barney got the first and only snow day of his career with the United States Post Office. He and Betty treated it like a holiday—they fed bread to the birds in their backyard and watched them come in flocks, and then they played a couple of games of pool, laughing and joking together. A few hours later, Barney died of a cerebral hemorrhage at the age of forty-six. Betty went on to become an enthusiastic central figure in the UFO community.

When she died in 2004 at the age of eighty-five, she had reported many more encounters, though the photographic and video evidence she offered was invariably inadequate. By then, her inner circle of supporters had slowly transitioned from a science-heavy group searching for truth, to starry-eyed believers seeking proof, and finding it all too often.

When Mike asked the state to approve the marker, he was asking them to cross one of two thresholds—approval would mean that either events of questionable proof should be included on the public record, or that the state considered the Hill case to be factual.

The stakes were also high for Mike; after a lifetime of quietly navigating the potential ridicule of being an abductee, he would be publicly associated with an alien abduction. But there was a compelling upside, too. The state's stamp of approval would create a New Hampshire culture that was just a bit more understanding of an untold number of experiencers.

One day, New Hampshire Public Radio listeners heard the station announce a new segment with a familiar musical bit—Mark Snow's "Materia Primoris," better known as the intro to the paranormal-themed 1990s procedural series *The X-Files*.

Then a voice came on.

"Hi. I'm Mike Stevens, of Farmington."

The host identified Mike as the man who had successfully lobbied the New Hampshire state government to, on July 11, 2011, erect a historical marker about Betty and Barney Hill. There it was, on Daniel Webster Highway, right next to recreations of a totem pole and a tipi on the grassy lawns of the Indian Head Resort. A critical development had come when Mike got support from Kathleen Marden, the Hills' niece. NHPR aside, most media reports focused on Marden, not even mentioning Mike.

"But he didn't care," said Andy, "because he wasn't doing it for the glory. He just really wanted to see the marker."

Without mentioning his own abduction experience, Mike told NHPR's listeners he'd had a lifelong interest in the Hills.

"I know that some people, 'cause of the subject matter, kind of think it's a joke, but every word on that was backed up by fact," Mike told NHPR's listeners.

And there it was, in state-sanctioned stone, stated as truth: the Hills "experienced a close encounter with an unidentified flying object and two hours of 'lost time' while driving south on Route 3 in Lincoln."

Val said that, other than his NHPR interview, Mike didn't talk much about the historical marker, or what it meant to him, personally. But Val sensed its importance. Without speaking, Mike glowed with pride.

And that was good to see.

What if there are aliens? Not just out there on some light-years-distant cluster of star systems, but here—in our solar system, in our skies, and sometimes, terrifyingly, in our bedrooms at night?

It's a powerful idea, one that could fundamentally alter our relationship with the universe. But this might play out differently for scientists and religious leaders.

Scientists have some amount of humility regarding our place in the boondocks of the Milky Way. UNH graduate Dr. Richard M. Linnehan, who maintains a queer double career of veterinarian and NASA astronaut, described the clumsiness with which we humans grope our way through space. He said that working on the outside of the International Space Station was more difficult than dog surgery.

"Picture yourself in a suit of armor on roller skates," Linnehan told the *New York Times*, "with a fishbowl over your head, wearing boxing gloves and holding on to a pair of lobster tongs, and trying to change spark plugs in your car in the dark."

Any alien that makes it to our neck of the universe will surely be vastly more adept at space travel, which implies an advanced level of technology.

Scientists can freely acknowledge this, and would use the encounter to bolster our storehouse of knowledge.

Churches, on the other hand, have little room in their dogma to cede their views of a higher power, which has led some devout to affirm that extraterrestrials are simply God's hitherto-unknown children. If that sounds like a fringe belief, think again: it's been affirmed by federal officials, including US representative Tim Burchett, of Tennessee. Burchett has consistently pushed the issue of UFOs in Congress, and believes there is an active government cover-up of contact between America and alien species. Speaking to a large audience on NewsNation, which regularly toggles between typical news subjects and UFOs, Burchett said that biblical references to a wheel of fire and angels actually describe spaceships and aliens.

"I think Ezekiel, I think he's talking about a flying saucer of sorts. . . . Acts, I've read that before and I've read different interpretations. And we were always taught to believe that those were angels and there was two of them," Burchett said. "I have no problem believing there's mentions of extraterrestrials in the Bible. That doesn't question my faith one bit. It doesn't hurt it, damage it. Matter of fact, it makes it stronger. It makes it far more vast."

George Coyne, Jesuit priest and former director of the Vatican Observatory, argues that Jesus died, not just for the sins of humankind, but for those of intelligent extraterrestrials as well—at least, those that are morally flawed enough to commit sins at all.

This is an effort to subsume the increasingly influential alien-centric culture into Christian teachings. But of course, the church seems to be at least as likely to be itself subsumed. If an advanced race of superbeings disembarks on planet Earth with nary a Bible tucked under their arms, do we really expect our relatively backwater theology to hold sway over that newly expanded reality?

Whether the aliens appear or not, various polls show that belief in their imminent arrival is growing, and many religion experts predict UFOs, not ghosts, will define the rising tide of paranormal spiritualism. Diana Pasulka, a religious studies professor at the University of North Carolina, writes in her 2019 book *American Cosmic* that alien beliefs are an emerging religion, one driven by influential Silicon Valley tech leaders and poised to overshadow current religious worldviews.

Whether our future culture is dominated by the specter of ghosts, aliens, or both, it seems like the landscape will be hostile to our current understanding of science, and of faith.

If, that is, any of those institutions remain.

After the Franklin Pierce Law Center found refuge under the umbrella of the University of New Hampshire, the rising tide of public distrust proved that no institution was safe.

In 2010, the state's electorate turned out a Democrat-controlled statehouse in favor of a veto-proof slate of Republican candidates who had come to see higher education as a low priority, or even a threat. At that time, per-capita support for higher education was $110 in New Hampshire, the lowest in the nation and far below the national average of $282. In 2011, the Republican majority enacted a budget that cut the state's higher-education funding almost in half, to a paltry $63 per capita.

Among the many reasons this was a terrible idea was one finding from the UNH Carsey School, which was continuing to explore the dynamics of public trust and civic health.

"Of all variables, education was proven to have the most relationships with civic health outcomes, indicating that education and civic health are critically intertwined," the Carsey School found, noting that this association persisted regardless of income and other demographic variables.

In other words, a state government concerned with dwindling public trust responded by strip-mining the one institution with a proven track record in building trust—college.

UNH's efforts to roll with the punches devolved into comic irony. As President Mark Huddleston announced the cuts, he said UNH needed a new paradigm to stay viable.

Though one of the chief values of public universities is their ability to provide services that transcend the demands of market capitalism, UNH began to aggressively monetize its archives of intellectual property. And it found that one of its most marketable properties was the Betty and Barney Hill Collection, which it promptly retrieved from the library basement at its Durham campus, and spit-polished to attract dollars.

Betty Hill had left UNH the star map she drew, her torn dress (with squares of fabric cut out for the testing), a sculpture made based on her description of the aliens, drawings of the aliens, and tons of papers—transcripts of interviews and therapy sessions, and Betty's journals. The collection also included recordings of Betty's hypnotic memory regressions (the value of which, remember, had been categorically debunked years ago by UNH hypnotism expert Ronald Shor).

UNH officials explained that the efforts of Mike Stevens had helped pave the way for the successful marketing of the Hill collection.

"The state put a historical marker at the site of the alleged abduction, so the significance of the collection in the public's mind and to popular culture was established pretty firmly," said UNH archivist Bill Ross.

Ideas are like viruses, pinging around the world and living or dying based on their perceived relation to reality. But as soon as an idea becomes monetized, it becomes a product. It's much less likely to die, because it causes money to change hands, and the recipient of that money will fight like hell to keep it alive, regardless of its relation to reality. UNH's stewardship of the Hill collection swiftly devolved into shameless cash grabs by schlocky partners. In the first of many UNH licensing deals, the Travel Channel show *Mysteries at the Museum* paid UNH for access to Betty's dress, which the show tested for irregularities. During the episode, a straight-faced lab tech asserted she had found "nothing to disprove" Betty's claims. Presumably, she also found nothing to disprove the existence of mermaids. Then again, in a world in which paranormal cable shows are underwriting New Hampshire's college degrees, who knows?

The Kitt Research Initiative, too, basked in the extraterrestrial heat. Andy and Beau had always been ghost-centric, but to everyone's surprise, the most well-attended regular KRI event proved to be Social Saucers, sometimes pulling in as many as thirty "experiencers" with firsthand knowledge of off-planet visitors. Attendees paid no gate fee, but were invited to donate. Some gave nothing. One attendee always put a check for one hundred dollars into the box.

The success of the event was largely due to Mike's personal qualities. He consistently articulated, day in and day out, for months and years, that his primary motivation was to help other traumatized experiencers.

Other UFO groups were "talking about it at the surface level," he said. "I really wanted to help the people who had been through this."

He began to see it as a mission.

"I feel like there was some sort of internal calling and push to do that kind of work."

During each weekly Social Saucers meeting, Mike would kick things off and then fade into the background, emerging only to offer a gentle word of encouragement, and to ensure that all those who wanted to share, could.

But on the inside, he himself was continuing to feel trauma from fresh experiences. The individual elements were difficult to decipher. White animals crossed his path in the woods, orbs of light floated into his room and resolved themselves into shadowy creatures, and once he'd glimpsed an alien that had cat eyes, but otherwise resembled the Creature from the Black Lagoon. He understood them to be screen memories installed by aliens who were initiating continued unwanted contact. But these experiences were offset in part by the upsides of Mike going public with his abductions. Family members told him that two years before his birth, his parents had pulled off the road and joined dozens of other motorists in gawking at a craft in the sky. His great-uncle had been expelled from a private New Hampshire academy in Wolfeboro after claiming to have been abducted. Others in his family had other striking experiences, which reduced Mike's sense of isolation. He speculated that his family—his bloodline—was of particular interest to the extraterrestrials.

KRI and UNH were on parallel tracks—their alien-themed offerings were bringing in money, testing both the large university and the tiny volunteer group, in their devotion to objective truth. Andy had drafted KRI's stated mission, to hold paranormal beliefs under a microscope, and screen out the falsehoods in the name of science. But when he sat in on the

Social Saucers meetings, he quickly determined that some of the attendees were cranks spouting obvious falsehoods. Screening the cranks out could disrupt Mike's dedication to providing a safe space for all self-identified experiencers; it could even disrupt the flow of donations.

They could not forecast how that action would play out. In this, KRI was an example wrapped in a microcosm inside a paradigm; but perhaps there was a key. That key was resolving a seemingly irresolvable conflict that would preserve two noble goals. How could KRI screen out falsehoods while supporting experiencers? How could UNH teach evidence-based science while monetizing unproven alien-abduction stories? How could the state of New Hampshire build civic trust while gutting its university system and bolstering belief in the Hills' spacecraft ride? How could the federal government regain the public's confidence when its members were supporting media outlets spouting dearly held conspiracy theories of hidden alien spacecraft locked away in military bases?

The monetization of aliens needed to be addressed for KRI—and for America.

Thirteen

SMOKE BREAK. SMOKE BREAK.

As a matter of fact we may in the dream ourselves live through almost all symptoms which we meet in the insane asylums.
—Sigmund Freud, *The Interpretation of Dreams*

There was a DVD player in the back of the limo. Andy had brought a selection of movies, and somehow *The Incredible Mr. Limpet* won out. He and Mike sat on bench seats, watching as a bespectacled and bow-tied Don Knotts fell into the ocean and was transformed into an animated fish.

Andy kept up a running commentary to Mike, who for the most part responded with noncommittal grunts. Andy said he wanted a cigarette. Mike did too. Andy knocked imperiously on the small window that connected to the front of the limousine.

The divider opened. Val was in the front passenger seat, and Beau was driving. She and her husband, Troy, had temporary use of the limo while a van their company owned was being repaired.

"Smoke break!" said Andy.

Beau was reluctant, but Andy had done the math. Fifty percent of the car's inhabitants smoked. And Beau used to smoke, he said. She should recognize the need to stop. Beau relented, and pulled over. Mike and Andy tumbled out gratefully, lighting up the instant their feet touched the blacktop.

The Ghost Lab

A short while later, the movie ended, and Andy watched the monotonous landscape of I-90 West roll by, dulled even further by the tint of the windows. He couldn't get the song out of his head. Not the whole song, really. Just that one line.

"I wish, I wish, I wish I were a fish," he sang. "'Cause fishes have a better life than people."

Val had started the ten-hour journey in the back, but migrated into the front. "Andy was driving us bonkers," she said.

Inevitably, he wanted another cigarette. Beau later imitated Andy's voice. "Smoke break," she said. "Smoke break. Smoke break."

Beau, exasperated, pulled off the highway. Again. She pulled into a big, empty parking lot.

"Val, are you ready for this?" she muttered. Val looked at her. "Okay, hang on."

Beau stepped on the gas and jerked the steering wheel to the right, hard, pulling the limo into a tight donut. Andy and Mike were briefly airborne as they crashed into the opposite wall of the vehicle.

Beau and Val squealed with laughter. Andy and Mike tumbled out of the vehicle, grabbing for cigarettes, cheerful and uncowed.

"Beau was not happy," Andy later said, magnanimously. "She was tired."

The laughter lifted Val's spirits. The most notable experience at her first ghost hunt had been freezing temperatures and a lack of bathroom facilities. This was her second, and it was already fun.

Sending Andy flying was a victory for Beau, but Andy was playing the long game. For years he would break up the silence of a room with the dread "I wish, I wish, I wish I were a fish . . ."

"Yeah, for all of Andy's ups and downs I never saw him as someone who worried about what other people thought of him," said Beau. She had come to think of him as a brother. "And I think that's a good quality of his."

Hours later, as dusk fell, they arrived in East Bethany, New York, and pulled into a long driveway to gape up at the looming wreck of the Rolling Hills Asylum, where Andy had in the summer of 2008 gone on a mostly-for-fun ghost hunt alongside OG television ghost hunters Jason Hawes and Grant Wilson.

Smoke Break. Smoke Break.

Today, KRI had a different goal in mind. They were at Rolling Hills on a secret mission.

To shut. It. Down.

A woman named Sharon met them at the front entrance. Andy was nervous. As Sharon reviewed the rules, Andy thought she had oddly emphasized certain phrases, perhaps even thrown a knowing look at Beau.

"Something she said made me wonder if she knew what was going on," said Andy.

Beau listened to Sharon's patter with a skeptical ear.

"She said, literally: 'We have a medium on staff whose job it is to keep them and hold them here,'" Beau later recounted. "They anchored spirits, the opposite of a clearing."

Beau was outraged. She leaned forward, but "Andy grabbed the back of my shirt and pulled me back."

"If you mess this up," he hissed into Beau's ear, "we're out."

"Fine," Beau said. She did what she always did. She pasted a smile on her face and nodded along.

The waiver asked them to agree to typical terms and conditions: no drinking, no open-toed shoes, no knives or pepper spray, no horseplay, no smoking, no outside food. But some clauses were atypical: no sage, no Ouija boards, and, importantly, no "Crossing over of Spirits, Exorcisms, Blessings, Cleansings, Clearings."

The KRI members signed it. But they were lying.

Beau and Mike slipped away to, they said, set up equipment, while Andy and Val told Sharon that the two of them would like the full tour. They wanted to distract her.

Sharon led them to a small one-room museum, which displayed several artifacts from the asylum's heyday, and then led them to the gift shop and eating area, where a staff member worked.

Sharon pointed out a book for sale about Rolling Hills history, and identified spirits the ghost hunters might encounter. There was a boy ghost named Little Jack who playfully said "boo," a cruel nurse named Emmie, an old woman named Hattie who continuously called "Hello!," and a seven-foot-tall man named Big Roy, who, according to the Rolling Hills promotional materials, "suffered from extreme gigantism, a physical deformity that left his face deformed, [and] his hands and feet oversized."

As Sharon ticked down the characters, Val heard a voice.

This lady is full of shit.

Val startled, but quickly noticed that no one else reacted to the words. She tried to hide her surprise. Nothing like this had ever happened to her before. "It wasn't my usual inner voice," she said.

Sharon told them that Rolling Hills was originally a stagecoach stop. It was converted into a poorhouse in the 1820s, and then an asylum. It later operated as an orphanage, a county nursing home, and a county hospital. In short, it was an institution that at one point had seemed as enduring as the sun. But eventually, Rolling Hills had gone sour with irrelevance. It passed through the hands of various private owners in the eighties and nineties, falling into disrepair. Eventually, it degraded into a heap of concrete bones, suitable only for those who wished to commune with the ghosts of what once was. Sharon called Rolling Hills the second-most haunted location in the United States (behind Gettysburg). There had been 1,700 deaths on the property.

Val heard the voice again.

She doesn't know what she's talking about.

In another cavernous wing of the asylum, Beau walked slowly but purposefully through the halls, hair upswept, carrying a set of big wooden beads that clacked with each step. Mike trailed behind with a video camera. A sense of fear permeated the place. It was oppressive.

They were here because of Beau's newly appointed spirit guide, Peter.

Peter had begun asking her to do things. It made no sense to her, but she always complied. "Like, I'd be driving and he'd be like, 'pull over' and just, you know, 'walk outside for me.' Then he'd tell me to 'get back in your car and go.'"

Smoke Break. Smoke Break.

Walking Elk, her prior spirit guide, had focused on Beau's role as a hollow bone between worlds. But Peter was more focused on helping spirits to move on. One day, Peter appeared to Beau in her bedroom.

"Listen," Peter said. "I need you to go to this building in New York." He didn't name it, but he said it was an old haunted medical facility. After Beau googled it, Rolling Hills came up, and Peter confirmed that it was the correct spot. Beau cruised the website.

"That was actually the first time I realized that people sold tickets for that kind of thing," said Beau. She was morally repulsed. The property owners were making money from trapped, suffering spirits.

"This," she said, "was a human zoo."

This was a similar situation to when UNH, and KRI itself, monetized the idea of aliens. But treating aliens as capitalist grist, at least, puts the main targets of the cash machine largely off-planet; in this case, Beau considered the targets to be a vulnerable population. Beau wanted to liberate every spirit on the property, and leave Rolling Hills bereft of ghosts. She went to Andy. If she signed up for a visit, she said, she ran the risk of the Rolling Hills owners looking her up, and realizing how much she emphasized the crossing of spirits.

"They're not gonna welcome me there if they know who I am," Beau told Andy. "I need a cover."

And so, they'd told Sharon that KRI was here on a ghost hunt.

Beau and Mike saw rooms in a state of intentional decrepitude, decorated with cast-offs and oddities. They saw a hospital bed, a straw-stuffed scarecrow wearing a Raggedy Ann face, and a necktie. One room had a dusty Christmas tree, complete with prop presents.

Beau followed Peter's instructions on clearing. In each room, she concentrated on the space itself, mentally filling each room with energy and then flushing it all away, like a psychic toilet.

"It was kind of like janitor work," Beau said. "Go in and clean this room, clean this room, clean this room, clean this room."

If Peter was her spiritual guide, Mike was her physical one. Every time she lost her bearings in the warren of rooms, he kept them moving in the right direction. They worked systematically. As each room was cleared, uncrossed spirits were pushed into rooms they had not yet visited.

After a few hours, the KRI group met outside for a short break, recorders still running. Andy and Mike lit up, and they all compared notes. Mike said he might have seen a shadow person. Val complained that she felt dizzy.

"I'm telling you," said Andy. "This place isn't nearly as bad as I thought it was going to be.... I gotta tell you, I'm actually kind of impressed that they're making an effort to capture the identities of the spirits of the people they believe are here."

Beau seemed annoyed, but was getting too tired to argue. Clearings were draining.

"Yeah, we'll talk about that," she said.

She disappeared into a portable toilet.

"That's going to show up on your audio, Beau," said Andy.

Beau sighed. "Yeah," she said wearily. "I know."

When she came out, Andy said that, when he was in the army in Alaska, he was taught to put his dirty socks on the toilet seat to prevent getting frozen to the surface, and then using them to wipe his ass. He threw the socks into the toilet afterward.

"Not a problem," said Beau. She pantomimed peeing from a half-squat.

They began to file back into the building.

"You know, I had a lot of wind earlier," Andy said. "But I kept breaking it." Everyone chuckled.

"That's on the recording," said Beau. Everyone laughed.

Beau and Mike continued the work. They entered a room with a low table full of ceramic knickknacks and boxes of dog biscuits. Beau crossed a little girl spirit. The next room held card tables littered with coffee mugs and vases and books and ice-cube trays. Cardboard boxes made half-hearted efforts to contain more artificial Christmas trees.

"Oh, crap," said Beau. "Someone had a baby in there, man."

Beau got upset at things unseen, grunting and gasping with effort and emotion. "Up and up and up," she muttered to herself at one point.

"We're almost done," said Mike, comforting her.

Beau sighed. "Okay." She stuck out her tongue in disgust, as if expelling something, and motioned vaguely at the air. "She was an angry person."

They rejoined Andy and Val in a room with paisley wallpaper that transitioned to small tiles near the floor.

"It looks like it was a bath that got wallpapered," Mike said.

"My stomach is starting to hurt again," said Val.

She walked around, sucking at her teeth in pain. "Oh. God. My stomach is pushing in on itself. Right here in the middle." She showed Andy. "Underneath my ribs."

They were nearly done. It was after 4 a.m., and the sun would be rising soon. Sharon came over.

Beau was bone-tired, but she tensed. Had Sharon heard her clearing a room?

The other group was not seeing many ghosts, Sharon said pointedly. In fact, they were going home. Sharon continued that she did not know why there would be so little activity. Beau glanced at Val.

"Oh, no!" Val said quickly. "I just saw a black shadow go down the hallway!"

"Oh, good good good," said Sharon (according to Beau). "I'm glad you're still having experiences."

When she left, Beau sagged, relieved.

"I remember being very thankful to Val for that," she said later. "She is so fuckin' smart, man. She didn't skip a beat."

Val had been complaining, but overall, the trip—the limo ride, the joint secret mission, the bamboozling of Sharon—had been a bonding experience for Beau and Val. Val had, for the first time, heard the voices that were a staple of Beau's psychic experiences. "I really started to understand that maybe I had a little more intuition than I thought I had," said Val.

The room they saved for last was at the far end of the third floor. This was the room that had given Andy a sense of unease during his earlier visit to Rolling Hills. From outside the room, Beau sensed a male spirit. Really, really intense, and very harmful.

"Okay," Beau remembered saying. "I'm going to do this one last spirit, and that will release everything."

"Can I go in and interact with the spirit first?" asked Andy.

Beau started to say no, but yielded. This was like a cigarette break all over again.

"Fine," said Beau. "But you know, protect yourself. Don't fuck around, man."

It was well known within KRI that Andy was something of a spiritual risk-taker. He rarely protected himself energetically, the way the others did. He himself didn't consider it to be a risk. He believed that his nature made him more or less impervious to the sorts of suggestive whispers that a malevolent ghost might make. But Beau felt that Andy's confidence made him a likelier target of malevolent spirits.

"I'll be fine," Andy said.

While they waited, Beau and Mike slumped onto a short table at the end of the hallway and watched as Andy and Val were reduced to shadows, walking down the dark hall and then disappearing into the doorway.

As Andy and Val entered, Val jerked her head to the left. She had sensed something in the corner. She could see nothing. But the room felt icky, somehow.

"Alright," said Val. "Let's move along."

They did some EVP (electronic voice phenomenon) work, asking questions into their recorders for later review. When they left the room, they walked back toward Beau and Mike, who were staring at them and whispering excitedly.

"Andy, are you on the left?" Mike called uncertainly.

"Yeah!" said Beau, happy for the validation of Mike's perception. "I'm seeing it too."

From their perspective, it seemed that Andy's silhouette was six inches taller than Val's. As they drew closer, and their features began to come into focus, the odd effect melted away. But Andy seemed off.

"Wow," Andy told Mike. "You're shorter than me."

He wasn't joking. He felt like he was looking down on Mike from a vantage point he did not possess.

"Whatever was in there hung onto him," said Val. She looked down into Andy's glassy eyes. Something was wrong.

"I feel like I'm on drugs," he said.

"Okay, babe," said Beau. Crises tended to put her at ease. She radiated a confidence that dispelled the fear of others. "We need to get this out of you."

And she did.

Smoke Break. Smoke Break.

As they climbed back into the limo van, they were confident that Rolling Hills Asylum was, for the first time, truly empty.

They had been awake for nearly twenty-four hours, but they had to put in another hour on the road. They finally pulled bleary-eyed into Andy's mother's driveway and piled inside, exhausted to the bone. Betty Kitt had Andy's restless intensity, but channeled it into hosting her son's friends. As they sat in the kitchen, she churned out plates of bacon, eggs, and toast, and set everyone up with a shot of liquor.

"You guys are gonna need this," Andy's mom said. "To help you sleep."

"And she was like, you're the medium. You're the medium," said Beau. "And she said something about her husband who had passed."

Betty was, Beau said, "sweet as hell," but it was 7 a.m. She was too exhausted to think about Peter Kitt's spirit, and she worried that Betty would begin to pursue a psychic reading like Andy pursued a cigarette.

"I'm just gonna go get something outta the limo," said Beau. Once there, she stretched out in the back and promptly fell asleep.

After they returned to New Hampshire, Beau heard through the grapevine of the spiritually attuned that in the wake of KRI's visit to Rolling Hills, the building had gone through an unusual period of senescence.

Some people made the connection to Beau, who had been open about the trip, if not her mission.

But Rolling Hills Asylum did not shut down. And people continued to have experiences there. The owners said Hattie, Little Jack, Emmie, Big Roy, and all the rest were still there. But Beau had an explanation. She said it was an entirely new set of spooks.

"Look around, there's an endless supply of 'em, and they're dying every day," said Beau. "So when they get stuck they wanna, you know, touch someone or do something like that, they're gonna go to places where people are actively going there to have that experience."

Beau was learning that spirits, once monetized, were almost impossible to dispel.

"They'll always be haunted," she said.

Fourteen

A WHISPER IN HER EAR

And if it were not for the rhythmic march
Of France and Piedmont's double hosts,
Should we hear the ghosts
Thrill through ruined aisle and arch,
Throb along the frescoed wall,
Whisper an oath by that divine
They left in picture, book, and stone,
That Italy is not dead at all?
—Elizabeth Barrett Browning, "Napoleon III. In Italy"

The twenty-two-year-old woman chugged along, south of Death Valley, through the deadly Mojave Desert along I-40 in her rickety little Honda Civic. She was in trouble.

She knew this without hearing those damnable voices in her head, the ones that sometimes issued her commands to do strange things. She usually did her best to suppress those voices, but now, against the ghastly silence of the desert, she would have welcomed even them. Her body was failing, and she didn't know what to do. So she kept driving.

The Ghost Lab

Something bad is going to happen. Because it's so hot. And I know that if I stop the car I will be dead. I won't survive it.

A stiff-scratched plate of lava fields topped with grit and sand, the Mojave is the most arid place in North America, so dry that regional rivers slip underground to pass, out of a healthy sense of self-preservation. A land's paranormal denizens are often refracted through the prism of geography. In New England, it was the cool mist of a gloomy, wet morning; in the Mojave, the paranormal appeared through a haze of heat lines that shimmered with ages current and forgotten—ghosts of Indigenous tribe members, of traders, of train robbers and sweat laborers; of military men testing Tiny Tim rockets; of gamblers who brought too much moxie and chutzpah to Las Vegas; of visionaries and cult leaders who sought to burn the rot of civilization and instead burned all they had; of hikers and endurance runners unprepared for the rattling pit viper or the great cats trained to bite at the base of the skull for the most effective stilling.

Haunted silver mines and ghost towns doubled as the stomping grounds of a series of improbable cryptids, including the nastily barbed (and notoriously drunken) Cactus Cats, the rifle-bending Yucca Men, and the Cement Monster, an eight-foot-tall Bigfoot-like creature that is occasionally glimpsed eating roadkill. If one could wring the glitter from the vast nighttime skies, a select few sparkling gems would prove to be alien crafts, heading to and from Area 51 in Roswell.

But these were not the Mojave dangers that weighed on the driving woman. Her concern was the June midday sun, which frequently roasted the surface to 130 degrees. The heat preyed on small errors of judgment. The yucca palms had dropped their annual flowers months ago, and only pointed at her accusingly with sword-shaped leaves. Her own mistake had been attempting to cross in her battered Civic, which lacked, among other amenities, air conditioning.

She gulped warm liquid from a gallon of water in her car, but it didn't seem to do any good. Rivers of sweat poured down her body, and she began to feel a bit sick, a bit woozy.

If her core temperature had already increased by just two degrees Fahrenheit, to 100.9, she would meet the clinical definition of heatstroke. She didn't have a thermometer, but it seemed she had surpassed that threshold long ago. The closer she got to 105 degrees, the more likely severe symptoms of delirium, seizure, and coma would presage a fatality.

She began to get tunnel vision. And she began to fade. Though she knew her situation was perilous, she couldn't prevent her mind from wandering. On some level, it made sense that this journey would be painful. It was an effort to transition away from the most difficult chapter of her life.

This was Antje Bourdages, in 1989.

Long before she turned to Beau for help with the mysteriously mobile Tupperware in her kitchen, even back when she was just a kid in Brooklyn, Antje was having otherworldly experiences. When she was a young child, voices began whispering in her ear. One day, her dog went missing.

Look under your bed, she heard, with sinister implication. The dog was indeed under her bed, and it was dead.

"It scared the bejesus out of me," Antje said. Because she was hearing things that others were not, she struggled to distinguish between her inner voice and disembodied voices. The busier and noisier her world was, the less likely the voices were to speak up. She grew to abhor silence, fearful of what might fill the void. And so she stuck around other people, or played music. "Bedtime was terrible. Nighttime was terrible," she said. "So yeah, it was pretty awful."

Shortly after she turned twelve, she told her mother about the voices. She wasn't sure what she was expecting, but her mother's response wasn't it.

Okay, she remembers her mother saying. *I think it's time to call the psychiatrist.*

Antje clammed up, and her mother eventually seemed to forget about it. The family moved to the New Hampshire Seacoast, where Antje went to Keene State College. There, she was in constant motion, pursuing the local party scene and drowning her eardrums in music and laughter.

After graduation, she and a boyfriend headed west in her old Corolla with few funds, and no sense of where they might land. When the road and money ran out, they were forced to rent the shittiest, seediest studio apartment on the Pacific Coast, where every click of the light switch was followed by the susurrant scrabble of a thousand tiny cockroach legs.

Poverty was a pressure cooker that steamed their relationship into something ugly and red. Her born-rich boyfriend had seemed breezy and worldly. But now, she realized he'd never had a job. All he could do was bus tables and wash dishes. They hocked whatever meager possessions the pawn shop would accept. Antje remembered a week in which they subsisted on a dollar loaf of bread and a jar of peanut butter, sitting on the seething Murphy bed to eat.

They began to bicker.

"Hey, we've got to get our shit together," she would say. "We can't do this." Every time, she had to force herself to say it, because she could feel him pushing her into a predetermined role of nag. He resented the confines of the menial jobs he was qualified for.

Antje landed an entry-level position at a securities and litigation law firm, running for coffee and taking messages. When she was promoted to a full-time receptionist role, clinging to every workday as a step out of the poorhouse, he got meaner than mean. One day at work, her coworkers noticed bruising on her arm. Over her protests (it was just an accident at the gym, she insisted) they pressed her until she revealed her legs and torso, also bruised. Only her face was unmarked, because when she had fallen to the ground, her huddled arms protected her head as he kicked the shit out of her. She was forced to admit that her life had somehow slid into an unrecognizable quagmire of domestic hatred.

Her employers did what she couldn't. They decided to send her to help a firm partner set up a satellite office out of state. That very day, they escorted her onto a train headed to the American Southwest with only the clothes on her back. And an old Walkman loaded with a single cassette tape: *Green*, by R.E.M.

"I could turn you inside out," Michael Stipe crooned in her ear as the landscape flashed by, endlessly, for twenty hours. ". . . Seems like you can't make it through."

Other than the fact that she wasn't being physically beaten, there wasn't a lot to be said for her time in the satellite office. All she knew about law was what she had picked up over the course of a few months as

A Whisper in Her Ear

a secretary. Her new boss was an impossibly demanding taskmaster. He pressed her to identify and negotiate office space, research legal cases, set up computer systems, and hire staff who were double or even triple her age, first interviewing them by quizzing them on their stances on arcane points of securities and fraud litigation, with which she herself had no experience.

In these pre-internet days, she would take a bus to a local public law school and spend hours fumbling her way through books of law, placing them side by side with a notebook and a legal dictionary, as if she was learning a second language by studying its most ancient and dry texts.

Every little stumble and misstep was met with derision and jeering by her new boss. The only way she could hope to meet his expectations was with endless stacks of eighteen-hour workdays. One day, she was five minutes late, and he lit into her so roughly that tears began to seep from the corners of her eyes.

Awwww, he said, in a horrific parody of empathy. *Does the baby need a widdle baby towel to cry in?*

The facet of Antje's personality that was ever submissive, ever eager to please, shattered. Furious, she spun on her heels and stormed out. She returned to the Pacific coast, working as a legal secretary for a different law company, one with a nice boss. For the first time, her life was looking up. When the Corolla died, she bought a used Honda Civic—the vehicle wasn't great, but the ability to choose and purchase it was. She was building a life.

Then a voice in her head ruined everything.

You need to go back home, it said. *Right now. You need to go.*

Antje didn't know why, or what, but it was so forceful that she felt she couldn't ignore it, not without bad things happening. And so she obeyed. Within days, she quit her job, got into the Honda Civic with no air conditioning, and headed east.

The voice didn't specify a route. So she chose to belt the country through the south, along Interstate 40. She could barely see the sweltering landscape anymore, between the advancing tunnel vision and the sweat pouring into her eyes. Had it sent her here to die?

At the last possible moment, she made out a modern-day oasis: a quirky building that reminded her overheated mind of the Bates Motel. She stopped the car. She hoisted herself out of the bucket seat and into those wavering lines of heat. The sun-baked ground rushed up to meet her, and someone switched off the lights.

She awoke in a convenience store adjacent to the hotel. It was ice. Ice! Bags of it, pressed into her neck and armpits, a shocking cold that acted as a welcome sink for the excess heat in her blood.

She couldn't speak yet, so people riffled through her purse, looking for her ID. Antje was twenty-two years old, of Keene, New Hampshire.

In one of those amazing coincidences that would eventually cause Antje to reject the notion of amazing coincidences, her saviors, who owned the hotel and convenience store, were originally from Keene themselves. They put her up in the hotel for two days, free of charge, and called her folks to let them know she was okay. They refused to let her leave, and so she became a sort of willing prisoner, well-fed and watered while she fully recovered. They relented on the third day. The third *night*, rather. While you're in the desert, they told her, only drive at night.

Several days later, she arrived home. Her sister took her out to visit somewhere in town, and pulled over on a sidewalk to say hi to a friend.

Hey, Charlie, her sister called.

Hey, Em, he called back.

As he walked away, Antje nudged her sister. "That's the man I'm going to marry," she said.

And she did. That same week, she was also hired by a law firm. There, she learned that the crucible of her legal field experience had left her with organizational and intellectual skills that belied her youth. She kept the voices at bay by fully engaging with the problems and tasks before her. She knew how to push past objections and navigate obstacles. She had become an organizational force of nature.

Over the next twenty years at the firm, the voices faded into distant memories as she wrangled a succession of complex and onerous clients through a complex and onerous legal system. She also used her skills to get through family problems, including an amicable divorce from Charlie in the late nineties. When she'd had the ghostly Tupperware encounter, she'd attacked the problem as if the culprit were roaches—she'd called a highly reputed professional, Beau, who had performed the equivalent of a psychic fumigation.

This was how Antje did things.

By the time she began hanging around KRI, she had been busy, by design, for decades. With a successful career, and the kids at college, it was finally time to relax a little. A new friend like Val helped her fill the void, and she had energy to spare to nourish that connection.

But the prospect of a quieted life carried a nagging strand of concern. What if, when things got too hushed, she heard a whisper in her ear?

Her entire life, Antje had treated the voices as if they were trespassers barging into her private thoughts. When she began taking KRI classes in 2011, Beau sent her to explore her inner terrain, which she was learning was as fierce and strange as the Mojave. She realized that the problem wasn't the voices. It was her response to them.

"I hate the cliché 'Take your power back,' but you know what I mean, to make it be okay for me," Antje told me. "To set boundaries so I wasn't at the mercy of it. That's terrible. That's when you get sent to the rubber room."

KRI also offered a new social life, even beyond Val, and Antje enjoyed getting to know everyone. Mike was unfailingly kind. She was impressed with Andy's scientific approach to the paranormal. He could be pedantic, but she sensed his caring, and appreciated that he would check in with her to ask about her day. Andy's tendency to goose the room with extreme statements created a fun dynamic—he was arrogant, but he left around himself a social space that allowed everyone to openly react to his arrogance.

"Mensa! Ohmygod, how many times did we hear that?" said Antje. "We would roll our eyes. 'Okay, Andy. You are the smartest person in the room. Okay. Got it.'"

As the months wore on, Antje sensed she was being assessed by the others. How she acted over the next few months would define her relationship with them.

"I don't think it was ever their intent for me to feel that way, but it was almost like I had to prove myself. I had to establish that I was reliable and I'm trustworthy and I really had a desire to bring my skill set to the Center," said Antje.

Antje was present at virtually every event KRI offered. She was one of the few nonmembers to join them on weekend hikes to local mountains. She gave generously to the donation box.

And so, when discussions turned toward possibly bringing in a new member to KRI, to help spread the burden of volunteerism-supported organizational tasks, talk quickly centered on Antje.

Beau was fervently in favor. Mike seemed to like the idea. Val couldn't get enough of her.

But Andy had one objection.

"I just did not like her," Andy said. This was in contrast to the impressions of both Antje and Val. "I never liked her. . . . She was like this really annoying puppy." Andy said he kept these feelings to himself. It was the sort of social lesson he had learned even before he stood before a mirror practicing how to say hello. "If they want to hug you, you hug them back."

Andy ultimately agreed to make Antje a member based on her kindness to Val. He called Antje her backbone.

"She helped Val in ways that no sane person would," Andy said. When Antje, whose law career was well-established, got wind of how difficult it was for Val to make ends meet one month, she gave Val a no-strings-attached gift of $1,000.

When they invited Antje to be a member, she was honored. She wanted to repay their offer of a unique life experience.

"I can only give what I perceive I have to give at the time and that was coming in and providing structure, making sure we were doing things the right way, the straight-arrow way that I learned how to do things," she said.

It was immediately clear to her that KRI was struggling to pay for its most basic needs; of the members, she was the only one with a stable, well-paying job. And so she started to add those little details that made KRI seem more official: she bought an exterior sign, and a key box. When it became clear that the events involving instructor-led meditation would benefit from more comfortable seating, she bought some beanbag chairs. She began to think about other ways to put KRI on the map.

"That was my offering. Financial offering, structure offering. That was my thing. But," she said, lowering her voice, "I think I did it so that people would like me. That's what I think. It was a way for me to fit in."

Fifteen
OUR ETHERIC BODIES

JUPITER.
No more, you petty spirits of region low,
Offend our hearing; hush! How dare you ghosts
Accuse the Thunderer whose bolt, you know,
Sky-planted, batters all rebelling coasts?
—Shakespeare, *Cymbeline*

When that midnight fire alarm sounded in that Syracuse hospital, something happened to Peter Kitt—something called death. But what is death, exactly?

In philosophy, the sorites paradox poses a deceptively simple question—if a million grains of sand constitute a heap, and you take away one grain at a time, at what point does it cease to be a heap? British philosopher Timothy Williamson has proposed that the sandpile is a heap inasmuch as people believe it to be so, which explains in one fell stroke the sorites paradox, and the reason philosophers are so rarely invited to dinner parties.

When Peter Kitt died, his bodily form was taken away, one dying cell at a time. At some point in this process (and the exact point is hotly contested), the subtraction of life processes created a situation in which Peter was no longer there. Death.

Scientists, the church, and spiritualists have starkly different ideas about what that means.

The large majority of people accept the spiritualist or religious proposition, that death, the natural process that has been studiously scrubbing every member of every species out of existence for the last 3.7 billion years, is not the final chapter of an individual's story. Instead, the person has a heapness that exists outside of the physical form, something so elusive to human measurement that the scientific establishment has utterly failed to document it in any meaningful way.

Sir Arthur Conan Doyle, creator of Sherlock Holmes and one of the leading Spiritualists of his day, asserted that spirits are humans that have cast off their bodies as we might cast off our clothes.

"Death changes nothing," Doyle said. "Our Etheric bodies are the same as our earthly bodies in all respects. Our Etheric mind and character is the same. We are the same persons but in another room."

And in Christian theology, our "Etheric bodies" persist forever in either agony or ecstasy, depending on the actions taken during a petty few years on Earth, most of which were spent as a mere child, or experiencing mental decline, or so beset with phone bills, the end of *Game of Thrones*, and other societal demands that one could hardly be said to be acting with the sort of perfect knowledge that ought to guide us through eternity.

One problem with this is that an increasingly robust body of scientific evidence shows that a person's identity is hopelessly entangled with the physical body—even in a theoretical post-flesh existence, the ghost would be so different from the person that linking the fate of one to the actions of the other would be grossly unfair.

For example, take the mortal sin of suicide, widely assumed to be a ticket to hell. Various studies have found links between suicide attempts and imbalances in the complex biome of organisms found in the human gut. Eternal damnation takes on a different framing if it is not the individual soul, but the soul's stomach contents, that is at fault for a mortal sin.

This is but one tiny teaspoon of annelid from a large can of worms. The gut holds sway over much of our behavior. One 2022 review published in the Cambridge University Press journal *Psychological Medicine* noted that the gut biome is associated with "several neuropsychiatric disorders, including major depressive disorder (MDD), bipolar disorder (BD) and schizophrenia-spectrum disorders."

Parallel fields of science show that other of our sometimes-sinful decisions are driven by hormones, chemical processes, environmental factors, and the traumas and joys of our infancy. Mike Stevens touched on this sometimes, during his Social Saucers meetings, when he described the impact of abduction trauma.

"Trauma physically changes the brain," Mike said during a speaking engagement. "The area of the brain that can help distinguish past and present memories shrinks, so you're almost constantly locked into that fear moment of the experience. No matter where you go. No matter what you do. Your brain always is in self-preservation mode. . . . You act differently. You're trying to process your own stuff. It breaks up marriages, destroys houses, friendships."

When the soul is unencumbered by such things, I assume that it gains a clarity of perception that elevates it to a higher state of being—but why, then, would there be a divine punishment for what it did when it was saddled with all of those factors over which it had no control?

This poses a bigger problem than evolution for Abrahamic religions, as well as for spiritualists who assume that spirits carry the same identity as they did in life.

But the scientists have their problems, too.

The scientific method—concocting hypotheses, conducting experiments, consuming data, and concluding conclusions—seems to be alive and well, and is achieving ever-more-defensible results. But if spiritualism and religion have a hard time conforming their worldview to facts, then scientists have a hard time conforming their facts to a worldview that anyone wants to hear.

Scientists have long grappled with how their findings are disseminated to the public, a process that academics call the "dominant model of science popularization." Ideally, the dominant model tiers information—it distinguishes minor scientific footnotes from the momentous findings that the public really ought to know about, and then distributes knowledge accordingly. But historically, our actual systems have never done that very well.

In the 1600s, when science first began to assert itself as the best prism through which to view the world, thinkers like Galileo and Kepler were entirely at the mercy of wealthy patrons who exerted veto power over their research. Communication to the (mostly illiterate) public at large was impractical and not a valued goal. The printing press created a means by which scientific papers could be mass-published for mass consumption by a mass audience, but for hundreds of years, this was largely hypothetical. Scientific knowledge was considered by those in power to be exclusively for those in power—the noble class and industrialists.

In 1774, when British chemist Joseph Priestley became the first to isolate oxygen, he advocated for the public, arguing that everyone, regardless of class, ought to have full access to the emerging body of science. This and other radical ideas were met with such fierce backlash that in 1791, a mob of the commoners for whom he advocated burned both his home and his church to the ground.

But Priestley's vision of a more democratic science became an inevitable reality, and by the 1800s, a science-hungry public was funding science programs through public institutions like governments and universities, allowing the first class of professional scientists to emerge. These scientists had much more academic freedom than their forebears, and they organized both to protect themselves from outside influence and to publish their work in direct communication with the public.

This was, perhaps, the ideal way to inform the public about important scientific discoveries. But it created new problems.

The increasingly jargon-laden papers scientists published were ripe targets for hijacking. Rather than being consumed directly by the public, they were first filtered through organizations with political agendas. For example, the Society for Promoting of Christian Knowledge and its ilk produced public education brochures that emphasized compatibility between current science research and a theological view of the world, while a variety of secular groups sought to do the opposite, tearing down religious views by reporting on evolution and other science findings as being at odds with a literal interpretation of the Bible.

For the science-minded, this was maddening. They had finally created a safe laboratory space for pure science, but it turned out that the world does not live in laboratories. No matter how much effort they spent to produce objective facts, those gleaming facts were instantly shattered on the rocks of subjective interpretation. Into the fray waded "honest broker" organizations and publications that sought to interpret science as the scientists intended, in a way that the public could appreciate. Enduring brands such as *Scientific American*, *Popular Science*, and *National Geographic* were all created in the 1800s as part of this trend. Their magazines used journalist translators to bridge the culture gap between scientists and laypeople.

Throughout the 1900s, the growth of educational and government institutions built a wall between scientists and the public. This allowed for a more objective and pure form of science, but at the expense of being seen as elitist or out of touch. And when the fruits of knowledge wreaked havoc on people and nature—nuclear bombs, pesticides, and lobotomies, to name but a few—the public wondered whether scientists on the other side of the institutional wall could be trusted.

Over the last two decades, the proliferation of the internet has allowed agenda-driven groups to once again overshadow professional science journalists. The scientific community has come to recognize this problem, thanks in part to UNH biologist Randy Olson, who in 2006 created a documentary called *Flock of Dodos* that highlighted the inability of scientists to convince the public that humans evolved from apes—even today, roughly 40 percent of Americans believe God created humans in their present form, sometime within the last ten thousand years. He suggested that without better communication skills, scientists would follow the dodo into extinction.

Narratives about death are just another example of this problem. Scientists have stacks of facts about what happens to our post-existence life force. If science's narrative of evolution fails to satisfy our need for greater meaning, then its three-word summary about death—we are gone—is even more deficient. If they could disseminate their multilayered narrative about the cycle of life and death in a different way, perhaps it could bring us all comfort, without compromising the existing evidence.

Science's story of death begins with the idea of energy, raining down like manna from our celestial sun and giving us life. We can extract star-blessed energy from our food only because we host a host of bacteria, viruses, and fungi that protect us from harm.

When we nurture a harmonious relationship with this literal multitude within us, we are repaid by humble microscopic workers—tiny planet-shaped *A. equolifasciens* toils, sucking soy into surface craters and expelling an isoflavone that reduces the risk of cancer and diabetes; *O. formigenes* looks like a mossy hot dog as it chips faithfully away at plant compounds that would otherwise give us painful kidney stones; and slipper-shaped *A. muciniphila* grazes on mucus, promoting a thickening of the stomach lining that helps to reduce obesity. Collectively, they contain hundreds of trillions of copies of DNA, which means our body is coursing with a blueprint for life that outnumbers our own by a factor of one hundred.

As living, conscious beings, most of us are takers, consuming huge amounts of natural resources to support ridiculously luxurious material lives. But when a person's lungs or heart cease their work, it is time for death. And death is a time for giving.

Our cells submit themselves to those tiny creatures over which we once held dominion. We trigger a chemical breakdown of our cell walls, and accumulating fluids flow downward, like tiny nourishing rivers. The residents of our gut biome eat energy-laden sugars, pushing into tissues and organs. The last parts of our body to be breached are the heart and the brain.

As the bacteria eat, they release ethereal gases—methane, ammonia, and hydrogen sulfide, which join beneath the skin. The air pushes up; the fluids push down, rending tissues asunder and creating new ways for our energy to be of service.

Bacteria convert our blood's oxygen packets into particles that glow with a soft, sea-turtle green. Tiny emeralds waft up into the world, announcing the commencement of a well-timed orchestra of bounty, and birth.

When they drift past flowers, they tickle the antennae of iridescent greenbottle flies. The flies abandon feasts of nectar and alight on us, their shiny bodies briefly holding our image as they lay eggs. Within hours, hundreds of thousands of tiny larvae hatch and race to absorb nutrients. They are gluttons, so eager to eat without interruption, they breathe out of their anuses.

Flies and wasps lay eggs that hatch into predatory maggots that feed on the greenbottle maggots. So do hister beetles, which creep out from underneath the corpse at night. So do skin beetles, mites, ants, and spiders. Birds and foxes, raccoons and snakes, frogs and lizards and turtles and salamanders, all eating the maggots that ate us.

Our energy has been multiplied, like biblical loaves, into a bounty.

Carrion beetles arrive, laden with mites, *P. necrophori*, that feed on the greenbottle eggs, which gives the carrion beetle parents a maggot-free space to raise their own young.

When our bodily fluids touch the soil, they awaken the bacteria *C. botulinum* and *B. anthracis*, which migrate up into the remains. If the drifting green sulfhemoglobin announced the beginning of the banquet, these two, which are responsible for the toxins botulinum and anthrax, toll a bell of closure, mediated by tolerance.

As the toxic bacteria spoil the meat, crows hurry to eat before decomposition goes too far, while fly species from the family Sarcophagidae will wait until the meat is so rotten that they can safely lay their eggs without worrying about predators.

The only animal that ignores this clock entirely is the mighty vulture, a symbol of hardiness and perseverance. Its bald head allows it to stick its beak into a corpse without getting feathers matted by infection-causing bacteria. The vulture urinates a strong acid down its own legs, purifying them by killing pathogens with a chemical fire. And its stomach is full of a gastric juice more powerful than battery acid, strong enough to break down bones, nails, and the toxic bacteria.

Finally, the frantic pace and competitive battles fade, and the place of our passing quiets. Peaceful mice and voles visit, chewing our bones to get a dose of calcium. Hushed tineid moths flutter down delicately to lay eggs; their caterpillars will eat our hair and nails. And some of the basic elements that were locked up inside us filter down to sanctify a sacred, rich loam. The grass grows thicker, and greener, creating a dense patch of foliage that marks our passing. We are here. And yet, we are everywhere, carried into the forests, the earth, the sky, the sea.

After a lifetime of disrupting Earth's natural cycles, we have paid a debt. We are free of responsibility, and engaged in absolution. Every measurable bit of our life force has gone toward restoration.

According to science, this is the story of our death. But no one is telling it.

Sixteen

IT SWEEPS YOUR DOUBTS ASIDE

> *How strange it is that any mortal in possession of his senses, should move a table, and not know it! And yet it is so, it has been so, but, we trust, it will be so no more. If any medium or tipper can gainsay this demonstration, we should be glad to hear from him, and would like to employ him, at a high salary, as a mechanical agent, to overcome for us, in a multitude of ways, the operations of gravity and friction.*
> —Charles Grafton Page, *Psychomancy: Spirit-Rappings and Table-Tippings Exposed*

As Antje began scouring listings on Craigslist and other online marketplaces, she had a mental checklist: a certain size, a certain height, and, critically, a certain number of legs.

Val had suggested the two of them devise a paranormal experiment, which dovetailed with Antje's broader campaign to advance KRI's mission. Antje had begun her time as a member undertaking a whirlwind top-to-bottom review that she hoped would turn the low-key, shaggy nonprofit into a sleek corporate show dog. She had the will, and the skill, to bring about legal harmony, and with that, a sense of professionalism and potentially new funding avenues.

Andy was fundamentally uncomfortable with following business requirements to the letter. It reeked of authoritarianism. Though Andy sometimes told me that he didn't like Antje from the start, other times, he moderated his tone a bit.

"It wasn't that I didn't like her. It's that she didn't want to do it the way I wanted it done. She is exactly who she says she is. A friggin' legal aide. Whatever. Arbiter."

Another time, Andy clarified that though he didn't like her, he didn't dislike her either. Despite participating in group norms like saying "love you" as a send-off, he seemed to think more about how her membership might impact the finances than anything else.

And KRI was humming along—it was hosting other paranormal groups from the region, and giving them advice on how to get more in line with scientific thinking. Andy developed a lecture that argued for the inclusion of mediums in ghost-hunting work, and detailed how to do so. With such exciting developments happening every day, the need to fill out forms was a dim and distant call.

"Andy would understand," Antje said. "He just didn't care." Using her knowledge of the legal field, she prowled through reams of papers, seeking the frightening specters of liability. A fire. An IRS audit. A lawsuit from a visitor tripping on the stairs. All of them, she reduced to mere paperwork.

Beneath her hands, the sorts of forms that were so off-putting to Andy were printed and shuffled and marked with the right arrangement of letters and numbers in the right color ink, and sent in envelopes to the world at large. Antje formed articles of incorporation and reviewed the leasing agreement. She undertook what amounted to a risk assessment, ticking off the boxes that, in the hands of her law firm's clients, had been left unchecked to disastrous consequence.

And now, she and Val were eager to make a meaningful contribution to the experimental side of the operation. After a few weeks, Antje finally found the table she was looking for. It was small, with three legs.

The three legs were important, according to Ginny May and Kate Otto. It's easier for a spirit to muster up the energy to destabilize three legs than four.

It Sweeps Your Doubts Aside

May and Otto had come to the Center last month, in January 2012, to put on a presentation of something called table tipping. They charged twenty dollars at the door, and got a full house.

May was a Boston-based registered nurse working in newborn intensive-care units, once called by a podcaster "among the best-smelling people I have ever known." Her casual childhood interest in the books of demonologists Ed and Lorraine Warren had taken a turn one night, when she felt an invisible presence sit on a bed next to her.

Otto had left behind a career in law-enforcement forensics for a career as a dog-trainer-slash-medium. She'd learned trance mediumship under Rita Berkowitz, a minister of Boston's Spiritualist Church and author of *The Complete Idiot's Guide to Communicating with Spirits*.

It was important that the two living participants of a table-tipping session like and trust one another, and indeed, in some ways, May and Otto were middle-aged, slim, casually dressed, feather-haired, cis female, Reiki-practicing peas in a pod.

With the crowd watching, they set up a small, wooden table about two feet wide on top, with three tightly clustered legs. May wore a long black scarf over a vibrant pink sweater; Otto had blue jeans over boots. They invited people to join them at the front of the room and examine the table for any hidden mechanisms. Then they sat on wheeled office chairs, occasionally reaching down to the floor to take a sip from their water bottles. They began asking the table questions.

Table tipping is a carryover from the Fox sisters–inspired séances of the late 1800s. Like spirit rapping, table tipping is an example of "physical mediumship," meaning that a spirit's responses can be heard or seen by everyone, not just mediums. It's like a Ouija board, except the whole table serves as the planchette, and there are no numbers or words. Beau had outlawed actual Ouija boards at KRI, on the grounds that the boards open a doorway to who-knows-what in who-knows-where for who-knows-why reasons. But table tipping occupied a sort of gray area, and so Beau sat near Andy in the back of the room, as curious as everyone else, and raised no objection.

May and Otto rested their fingertips atop the table, very lightly. After their first question, the table was still for a moment. Then it began to lean, first to one side, then coming back to the ground and carrying that added momentum to lean up further on the other side. May and Otto kept their fingers in contact, but insisted that the table was entirely responsible for the movement.

As the table warmed up, it rocked, it rolled, it careened back and forth between the two women in irregular, organic arcs. Antje noted that their hands were visible, and their feet were flat on the ground. It didn't look like they were applying any pressure at all, certainly not enough to produce the extremities of movement the table was displaying.

"You could see their knees underneath the table," said Antje, ". . . so there was no, there was no possibility that anybody was kicking anything."

When the women asked the name of a spirit in attendance, the table rocked as they chanted letters, *A, B, C, D, E* . . . The first letter was *M*. Then they would start from *A* all over again to get the second letter.

"To me, they seemed legitimate," said Val. "It seemed like they were getting answers."

Beau could, of course, see and hear the spirits at the table, spelling out answers in the tortuous, letter-by-letter process.

"Wow," Beau thought snarkily. "This is, this is ineffective. Like, this is not very efficient."

A, B, C, D, E, F, G, H, I . . . The second letter was *I*!

Beau glanced at the time impatiently.

"The dude's name is Michael!" she muttered to herself. "Come on!"

A, B, C . . . The structure of the session followed a protocol established by paranormal researcher Kenneth Batcheldor, a British clinical psychologist who conducted a series of table-tipping experiments in the 1960s and 1970s. As with hypnosis and mediumship, Batcheldor concluded that table tipping was successful when it was done by people who believed that it was possible.

In other words, table tipping is a bit like government—it can function only with trust.

"In most cases the table will start to move due to unconscious muscular activity. This can give an amazing illusion that the table is moving of its own accord as if animated by some mysterious force," wrote Batcheldor.

He was skeptical of spirits, but ascribed the motion to a different paranormal force, telekinesis. He believed the sitters were moving the tables with their minds.

"You get the impression you are already succeeding in generating paranormal movements. This has precisely the same impact on you as real success would have—it sweeps your doubts aside and produces total faith," wrote Batcheldor.

While table tipping at KRI, Otto told the audience that her daughter had died, and was among the spirits animating the table. At one point, the table leapt up onto its side, pressing one edge against Otto.

"That's my daughter giving me a hug," she told the audience.

After a few weeks, Antje managed to snag the perfect table—small, lightweight, and just the right height. She and Val wanted to learn how to tip tables, a first step in forming a KRI-affiliated group of sitters, and eventually standardizing a process that would be replicable in a laboratory setting. To ensure that they would have a baseline of factors to tweak, they developed a protocol on hand and foot placement, spiritual intent, and distance from one another, as well as a regimen of questions to be answered.

When everything was in place, Antje and Val took the table into a shared workspace and sat opposite one another. They placed their hands gently on the table, lying flat. By prior agreement, their feet were tucked back, and flat on the floor. They were alone.

Antje began.

"Is there an energy here that would like to communicate?"

Antje hated that question, even though she asked it all the time.

"It drives me crazy when people ask that question. It's dumb. Do you know what I mean?" Antje said. "It's a dumb question. But we had to open the door somehow."

The table remained still as stone. They sensed, intuitively, that a spirit was present, but it could not be induced to budge the table. Not even a little bit.

A few days later, Val and Antje tried again. And again. Over the course of the next couple of weeks, they tried six more times. The table never so much as trembled.

"It was kind of a bust," said Val.

One day, Val couldn't make it. So Antje asked another developing KRI medium named Shannon to fill in. After following the protocol, Antje again asked the first, dumb question. And the table exploded to life beneath her hands, rocking as enthusiastically as a wriggly puppy.

"It was, it was insane," said Antje. "I have never experienced anything like that before."

They ran and got Andy. As he watched, looking for mundane explanations, the table once again hurtled obligingly around the space. When Beau came to observe, she saw a blue light emanating from the base of the table. In its glow, various spirits were lined up to take a turn. Beau realized that, for them, it was fun. Though it was less efficient than a simple conversation, they liked batting the table around.

"It was the enjoyment of interacting with this reality," said Antje. The message was the medium, one might say.

During her next session with Val, Antje was excited to see whether they would have success. But again, they had nothing. Val tried with Shannon, to no effect. Finally, they decided to try with three sitters. Val, Antje, and Shannon all gathered around, reset their intentions, and placed their hands lightly on the table.

Bupkus.

Then Shannon and Antje tried again, and again, the table responded to questions by moving animatedly beneath their hands. Antje was flushed with the thrill of it. But as she began talking about the next steps to take, she noticed Val, watching, an inscrutable expression on her face.

"I think Val felt a little bit left out," Antje said. Was Val jealous? Antje considered asking Val about it to clear the air. What she really wanted to say was that her success with Shannon didn't have anything to do with Val. That was just the way the table tipped. "It doesn't mean anything about anything."

But Antje worried that broaching the topic would make it worse.

"Making it more of a big deal out of it would cause strife, you know what I mean?" she said.

But it took the joy out of tipping the table, which had, after all, been Val's idea in the first place. "It was just, it was an awkward situation," said Antje.

Very soon after, she put the tipping table into a closet, and forgot about it. She and Val had lost the appetite to do public performances, or form a sitter's group, or even to further study the phenomenon.

"And so that, that was the, the end of the table tipping," said Antje. "It just kind of faded. Because of everything."

Not long after, Val called Antje. Antje didn't answer, so she left a message. For the first time in their friendship, Antje didn't call her back.

It was 1925, and Sir Arthur Conan Doyle, sixty-six, took the stage, looking out at a boisterous crowd dominated by Spiritualists. Doyle's entire lifetime had taken place in a world influenced by the Fox sisters and their followers, and he had become one of Spiritualism's leading public advocates.

Having recently been named honorary president of the International Spiritualist Congress, he was in the Salle des Sociétés Savantes, a hall on rue Danton in Paris, to give an address on the wall-rapping, spirit-writing, table-tipping, orb-photographing movement.

Four thousand people had crammed into the hall, with 1,500 more smashing barriers on the street and drawing beatings from police in their bid to gain entrance.

Just as Doyle's wife, Jean, moved to turn on the lights and tell the doormen to stop admitting people, the doors broke inward and the crowd spilled in, knocking her aside. Fights broke out. A heckler shouted an ectoplasm joke, and a Spiritualist broke a chair over the head of a skeptic.

Doyle stormed off the stage. When order was restored, he returned, holding a cane and his notes. The French phrases that emitted from beneath his great walrus mustache were made new by his hearty Scottish accent.

"I am proud to speak in the intellectual center of the world; addressing myself no longer to unbelievers and sceptics, but to brothers in belief, our allies in the great war which we Spiritualists are all waging; the war against materialism," Doyle said.

Doyle's speech was not simply an affirmation of spirituality; his address would have likely qualified as treasonous, were it directed at his government, rather than materialism generally.

"This is the most terrible enemy; it is the cause of all the evils of humanity," Doyle thundered. "It will destroy the world if we do not succeed in strangling it, in overthrowing it. It presents itself as a many headed monster, the materialism of a society thirsting after pleasure and material riches, the materialism of science, materialism of morals, materialism of Churches, temples, mosques and synagogues, formalist, dogmatic, and trivial, forgetful of that which is true living communion with the Higher World."

After these stirring words, Doyle then moved on to the substance of his remarks, which was to offer arguments on the reality and nature of ectoplasm. Sadly, it was not Doyle's night. The projectionist was either drunk, a willful saboteur, woefully incompetent, or all three; images meant to illuminate his discussion on ectoplasm arrived far too late, far too early, or upside down, emboldening the heavily outnumbered skeptics in the audience to laugh and jeer.

But Doyle's opening remarks had already set the tone. As Spiritualism ascended from the cradle wrought by the Fox sisters, the community's ambitions evolved from simply existing alongside the dominant paradigm, to upending it. When a belief in spirits can lead the creator of Sherlock Holmes to declare war on the social order of the day, it becomes important to understand how such movements grow into institutional threats.

Certainly, it's not the factual evidence underpinning the belief system. Doyle gave his address to a record number of believers, thirty-six years after the Fox sisters admitted that they'd inadvertently begun the Spiritualism movement as a toe-cracking prank on their mother.

When and why do people begin distrusting institutions and searching for alternative views of the universe? After studying the relationship between trust and major global events, experts at the Ford School of

Public Policy of the University of Michigan concluded that the institutions themselves bear much of the responsibility.

Robert Putnam, one of the world's leading experts on public trust, happens to live in New Hampshire. "Trusting somebody who's untrustworthy is not a virtue. That's gullibility," said Putnam, a professor of public policy at Harvard University's Kennedy School of Government, in a public radio interview. "And so the question is, is this a trust recession or a trustworthiness recession?" Political institutions tend to make decisions that mediate two forces—science, and lobbyists representing wealthy corporate interests. That freezes out an important stakeholder—local communities, which bring their own unique community expertise. One example of the shortsightedness of this approach came in Flint, Michigan, in 2014, when public officials ignored water quality and health complaints of residents until the environmental-justice travesty became national news. As a result, trust in government certainly suffered.

May and Otto had taught that, in order for table tipping to work, partners had to trust and like one another. It's clear that trust and affection are also important for an institution and the population it serves.

Institutional fiascoes such as political scandals, economic recessions, the widespread failure of the Texas power grid in 2021, and the systemic racial bias of America's justice system all cause distrust to spike. I saw this while living in Chicago in 2000, when George Ryan, the Republican governor of Illinois, who had voted to implement the death penalty in 1977, announced that he was commuting the sentences of all death row prisoners. He did this based on findings that a majority of the convicted inmates were innocent of the capital crimes for which they were sentenced to die. He and I were staggered at the evidence that the institution of law in the state was so horrifically flawed. The taint of that scandal has colored my perceptions of American justice ever since.

And so, if an institution wants to be trusted more than a regional ghost-hunting group, it must earn trust through competence, but this is even more difficult than it sounds. Institutional trust takes a long time to build, and a short time to collapse, according to David King and Zachary Karabell, authors of *The Generation of Trust*. They found that institutions need to publicly demonstrate that the institution and its leaders are acting with integrity to accomplish a respected mission.

American institutions are not at all well-positioned to make up ground against paranormal thought-leaders. Though one can point to various broad societal successes in the modern American age—for example, violent crime fell by 50 percent between 1993 and 2022—there are fundamental institutional failures that have provided a tragic backdrop to the modern growth of spiritualism and paranormal beliefs. Here, in the most technologically advanced and wealthy country in the world, income growth has stagnated, income inequality has skyrocketed, and the average life expectancy has declined.

Learning about the knottiness of rebuilding public trust led me to a fairly simple, and terrifying, logic proof, using the sort of "if-then" statements one finds in a high school textbook: if institutions need public trust to survive, and if institutions need to perform better to earn public trust, and if institutions are too dysfunctional to perform better, then . . . institutions are too dysfunctional to survive.

Seventeen

ARE THEY GETTING FRESH?

> *"Lead on!" laughed Chet. "It'll be a brave ghost that will tackle the whole five of us."*
> —Franklin W. Dixon, *The House on the Cliff*

As Val walked down the old button factory's unfinished hallway, Beau trailed behind, filming with a handheld recorder.

"DA-da . . . DA-da." Val was singing the famous score from *Jaws*, but ended it with an odd little flourishing "Da-daaaah!" that turned it into a show tune. She seemed almost giddy. After years of standing alongside Beau's casual spiritual brilliance, this was a rare opportunity for Val to shine. It was late February 2012, and already, ocean-generated warm winds were taking bites out of the Seacoast's snowbanks.

Beau had, after months of teaching both Val and Antje, declared that they were far enough along in their mediumship to help KRI tackle a long-anticipated experimental method: the triple-medium walkthrough. They were seeking shared visions—otherworldly knowledge that could be confirmed independently. Establishing a successful protocol could produce hard data—What sorts of observations tended to be hits? Which mediums made those observations? How often did hits occur? Andy, who had been recently pulled over for speeding by a big, beefy Irish cop from Greenland, was eager for some good news.

This was also one of the first investigations at which KRI had shown up in full force: Andy, the mad scientist and leader. Mike, the quietly compassionate and hulking techie. Antje, the by-the-book intuitive. Val, the sensitive empath. And Beau, the undisputed queen of mediumship.

Val and Beau followed the hallway into a cavernous space where large pieces of machinery clogged the floor. As soon as Val entered the room, her cheeriness faded. She stopped and looked at Beau with something resembling distress. She tapped her chest.

"I can't catch my breath," she said.

The building they had chosen for the triple-medium walkthrough was the Button Factory in Portsmouth, which was quite a pile of bricks. In fact, the bricks that made up the walls had been purchased and delivered to a field next to Islington Creek in 1895, even before the Portsmouth aldermen had approved it. Various haphazard additions had caused the original factory building to sprout annexes, a third floor, wings, and staircases.

When the market went elsewhere, all that remained was an empty industrial warren crisscrossed by smokestack shadows. The spaces KRI was exploring had been converted into artist studios littered with remnants of oddball artistic projects—one renter restored stained glass, another made functional outdoor couches from grasses planted on earthen berms, and a third painted canvases that were meant to be placed on the floor.

For a while, in the late 1980s, the Button Factory also hosted another tenant: an office for a local architect—Antje's father. Her memories of him saying the building was haunted had first motivated her to contact the owners and set up the current ghost hunt.

Antje's excitement over this investigation had manifested in a frenzy of organization. She had volunteered her SUV to carry most of the team, and all of the camera equipment—which this night included seven audio recorders and six cameras. She'd called the owner and procured access to a communal woodworking studio, a pottery space, various offices, and some creepy restrooms, all located within the first floor of one wing of the building. She had asked the various renters to sign waivers agreeing to let KRI into their particular corners. The hallway light had to stay on, she noted, because the fire department considered it a liability otherwise.

As they began to set up the cameras, Val and Mike unexpectedly saw a broad-shouldered artist in painter's whites walking away from them, down the hall. It was unclear whether he knew they were there, so they followed him, to clarify the purpose of their presence. He took a left turn at the end of the hall, and when they arrived, they saw that there was no left turn to take. He had vanished into a blank wall.

Though they hadn't caught it on camera, their first sighting of a full-body apparition had left them as exhilarated as a birder stumbling across a pink-headed duck. The Button Factory was certainly haunted. Now it was up to the triple-medium walkthrough to learn more about the spirits who lived there.

First, Beau walked through the woodshop on her own. Eventually, the protocol would require complete separation of the mediums, but for now, they were just seeking affirmation that shared experiences were possible. And so, Beau was now trying not to let the details of her experience slip as she filmed. Val described the sensations that accompanied the chest pain she felt upon entering the woodshop.

"Immediate discomfort. . . . Dizziness. Really bad dizziness. . . . Nausea," Val said. "My hands feel like they're going numb."

"Am I allowed to say anything or no?" asked Beau.

"I think you should," said Val.

"'Cause that strikes me as blood loss," said Beau.

"Yeah," said Val. "That makes sense."

Val sounded a bit deflated. Though she was now a full-fledged empathic medium, she was continuously in deference to Beau's expertise. She both appreciated her mentor's presence and saw that it was an obstacle to coming into her own as an intuitive.

"I'm still learning," she said to Beau. "So anything you can add is helpful. I don't feel anything strongly in terms of individual spirits. I'm just feeling strong feelings. But," she added, dropping her voice down to a whisper, "you never know."

Val stood in front of a gray plastic trash can with boards sticking up out of it. Val now sensed four males, one of whom she described as cartoonishly evil. She imitated him, rubbing her hands and licking her lips and leering.

"I kind of think that's the guy that Mike and I saw," she said.

"There's definitely a main man," said Beau. What Beau did not reveal to Val was that she had seen a spirit who had, in life, worked here, sexually assaulting the female factory workers with impunity.

Val drifted to a new area of the room. There were fire extinguishers mounted on metal support columns, and workbenches littered with equipment and what might have been the ribs of a keel. One of the renters was a woodworker who made small traditional boats.

"Of the whole room," Beau told her, "this is where I would stand. To do the work."

"Yeah, I can see why," said Val. "Yeah, I'm getting that blood loss feeling in my arms again."

"Are you willing to try something?" asked Beau. "Are you willing to lay down on the floor?"

Soon, Val was flat on her back, hands resting on her stomach. Stools and sawhorses loomed over her. She wore dark jeans and a heavy green jacket with a furry frill.

"I can feel this one person standing over me," she said. There was a long pause.

"Just know that you're safe, and ask them to do what they wish," Beau said.

"Hmmm. Are they getting fresh?" Val asked. "I'm feeling something, like on my inner thighs." She chuckled.

"You can ask them to fast-forward," said Beau, referencing a training exercise in which mediums could control the scenarios played out by residual spirits. "But. That makes sense."

"I feel it in my arms," said Val. "My stomach is just throbbing."

There was another long pause. Then Val asked tentatively.

"Did this guy like, rape and mutilate women?"

"Mmm-hmm," said Beau. This was a staple vocalization from Beau. It affirmed the experience, but she delivered it almost chirpily, to reassure Val. One of her core teachings was that entities in the spirit world could only hurt the living if the living allowed it to happen.

"Choke?" asked Val.

"Mm-hmm."

Val was still except for her fingers, which occasionally drummed on her stomach.

"Hmm," Val said. "That feels funky. I'm feeling like everything is being sucked down into the floor."

Then, a moment later: "Okay, that's like really—I'm really feeling that. That's not like, sixth-sense feeling. That's like, he's physically touching me."

When Beau didn't reply, Val joked, "That's the most action I've gotten in a while."

Beau told Val to be defiant with phrases like, "You're nothing, you're dirt. Little man, you can't hurt me."

"He's not wasting any time," said Val. There was another long pause.

"You okay there?" asked Beau.

"Other than feeling violated by a ghost, I'm feeling good," Val answered.

Val's arms were now spread on the ground, near her side, palms down. She said that, when she tried to fast-forward the spirit through the act, the spirit resisted. Beau addressed the spirit directly.

"Okay, so you're done. You're finished," she told him, in a tone that brooked no nonsense. "And now she's going to tell. She's going to walk right out these doors and tell everybody what you did."

Val's fingers beat on her stomach again. Then she sat up, running her fingers briefly down across her pubis, brushing something invisible away, as she pulled into a sitting position.

"I think I'm done with this exercise," she said, a little whispery, perhaps a little upset.

"Okay. Sure," said Beau.

"I really feel like I had sex," said Val.

"Yup," said Beau. "Sorry."

"That's okay," said Val. "I really didn't think that was po—"

"Possible?" Beau finished. "Yeah, it is. There are ones that are way stronger than what we normally run into."

"I feel like I'm getting pulled back," said Val.

"Mm-hmm. Tell him," said Beau. "You can't come back."

"Nope, I'm good," Val told the ghost rapist. Adding, somewhat absurdly, as if she were rejecting a clumsy nightclub advance, "Thanks anyway."

The other team members were also making their way through the Button Factory. Occasionally, they talked, but no one whispered. Andy had banned whispering, because it made it harder to tell whether the resultant audio was a team member or a ghost. If someone accidentally whispered, or produced a creaking sound by stepping on a floorboard, they would say, "That was me."

This was one example of how Andy's protocols had advanced significantly since the Rolling Hills Asylum adventure.

Each investigation started with a checklist—did they have batteries, and fully powered devices, and working backup voice recorders, and power cables? They wore identical voice recorders on black armbands. They did an equipment check, to ensure all settings on devices were correct. Then they would stand in a circle while Andy clapped his hands, so that the audio could later be synced perfectly.

After Val's experience, it was Antje's turn for a walkthrough. But Mike and Andy needed to first do some camera work in the woodshop, so the women chatted, with Beau and Val studiously avoiding the topic of the ravenous ghost.

Beau told a story about refinishing a refrigerator without a mask on. "Dude. I'm not going to breathe through my mouth," she chuckled. "Refrigerator paint on every nose hair. I had to remove every one of my nose hairs. And they were hard. It was painful. I should have taken a picture."

Are They Getting Fresh?

Val and Antje laughed, but at times, it seemed like a conversation between Beau and Antje, one that Val was simply listening in on. When Antje stopped answering phone calls from Val, it was part of a broader pattern of distancing herself from Val. She was still friendly on a surface level, but there was no more hanging out, or references to them being soul sisters.

"I was kind of hurt by that," Val said later. She felt Antje had simply switched her out in favor of Beau. "It seemed to go from, Antje just seemed to go from me to Beau kind of like a light switch," said Val. "She was calling Beau, hanging out with her before and after events."

Antje and Val never talked about the change, but Antje later told me that she had begun to see Val differently. The more Val opened up to her, the more she began to see Val as being enmeshed in her victimhood.

"It became a little bit difficult to be around," Antje said. "Because it was always kind of heavy. She was hypervigilant. . . . When something good happened to someone else, she would appear to feel put out. She would appear to feel, not jealous, but: 'Well, that kind of stuff never happens to me.'"

Antje didn't like feeling as if she had to navigate this side of Val, to tiptoe around her own happiness.

"You go to your friends to express joy or to share in an exciting thing," Antje said, "and I could never really share something good."

As the three women waited to reenter the woodshop, Antje topped Beau's painted nose hairs story. Roughly fifteen years ago, she said, she'd fielded a phone call to the law firm from a man who masturbated against a belt sander in a woodshop during lunch break and slipped, ripping himself "from stem to stern." He'd stapled himself back together with a staple gun.

"I'm wondering if I have a claim," she remembered him saying. "Because I'm really injured."

"Dude!" said Beau.

Val cut off the conversation.

"I just saw something out of the corner of my eye that way," she said. Beau and Antje looked, but saw nothing.

Finally, Antje entered the woodshop. Beau once again carried the camera. Val stood quietly in the fringes.

As soon as she entered the room, Antje reported physical symptoms. Her nervousness at being center stage flowed out of her as rapid-fire speech.

"I feel pressure in my chest. As soon as I walked in here, it feels different from in that hall," said Antje. "I think of someone dying from a heart-related thing. That's how my dad died. That's how I interpret that feeling. Whenever he's around I kind of get that sensation."

Antje said she sensed a man in the corner. He was older, with a long, gray beard, grimacing at her with crossed arms. He didn't necessarily strike her as dangerous. After a moment, Beau repeated the request she had made to Val.

"Lay down. If you don't mind laying in the dust," she said.

By the time the words were out of Beau's mouth, Antje was on the ground.

"See how I don't ask questions," Antje said, looking up and grinning.

As Beau changed position to get a better camera angle, Antje reported her first sensation.

"I almost feel like I'm on a pitching ship," said Antje.

"On a pitching ship?" asked Beau.

"Yup," said Antje. She moved her arms up and down in an alternating pattern. "I feel the pitching." About twenty seconds passed.

"Why do I feel like someone wants to, uh," Antje chuckled. "I don't want to say this."

"Probably because they do, sweetie," said Beau. "Go with your comfort level."

Antje laughed uncomfortably.

"I feel like someone is saying, 'Spread your legs, bitch.'"

"Okay," said Beau.

"I feel like someone is looking at me with hunger. A mean hunger though. Ugh."

"Only go to your comfort level," said Beau.

"I feel very vulnerable down here," said Antje.

"See what you can—"

Antje cut Beau off, speaking quickly.

"I feel like at first I was being lulled into a false sense of complacency. This person was being false. Like they put on a very good face, like a very good exterior. But they had something mean and nasty behind it. 'I'm gonna take you and you're never going to be the same again. And you're going to like it.' Oooh, I don't like that. Uck."

Antje rubbed her pubis downward, the same gesture that Val had done a short while ago.

"I feel like there's someone right here," she said, gesturing to a space inches from her face. "He's surprised that I'm not afraid. Confused that I'm not afraid."

"Ask him what happens if you tell the authorities," said Beau.

"He's laughing," said Antje. She related his words. "They won't do anything. I'm a pillar. Everybody loves me."

At Beau's direction, Antje threatened to expose him, and then he backed away, fading from the edges of her perceptions. She got up quickly.

"Did you get the same thing?" she asked.

Beau laughed.

"Yeah," said Val. She and Beau vigorously brushed the dust from Antje's clothes. "He was touching my thighs. Bad things."

In such a moment, the tensions between Val, Antje, and Beau melted away.

For these three women, who had all been hurt at some point in their lives by men, the experience would have qualified as therapeutic for practitioners of psychodrama and similar trauma treatments involving group-supported roleplay. Here, while standing in sisterly solidarity, they'd used empowered thoughts to confront an abuser who was only his worst sins, as hateful as he was insubstantial.

After parrying his assaults, the three women worked together to send the spirit on, but he clung to our plane like a barnacle. Val distracted him, while Beau and Antje worked to immobilize him.

Antje's spirit guide told her precisely where to pinch the air, and just how to pull the things she could not see, while Beau did the same thing on the other side. In this way, they wrapped the rapist ghost up until he could not move. They had broken him. In the blink of an eye, he was banished, forever.

Here, in the bowels of a run-down factory, over the five-hour investigation, they'd engineered a spiritual retributive vigilantism in which the victim was empowered as judge and executioner, the accused had no right to defense, and the sentence eliminated all future risk. It was an individualistic, experience-based form of justice that was everything our institutional justice system is not—simple, functional, and satisfying to the core of their beings.

Eighteen

A CRANE, A LEMUR, AN OCELOT

> *St. Michael the Archangel, defend us in battle. Be our defense against the wickedness and snares of the Devil. May God rebuke him, we humbly pray, and do thou, O Prince of the heavenly hosts, by the power of God, thrust into hell Satan, and all the evil spirits, who prowl about the world seeking the ruin of souls. Amen.*
> —Prayer to Saint Michael the Archangel, patron saint of police, Chicago Police Chaplains Ministry

In 1900, a cowboy-turned-penmanship-scholar named Joseph Hesser opened a small private college with a humble mission of boosting the downtrodden into the middle class. For one hundred years, Hesser College helped nontraditional students attain dependable paychecks in decent careers: preschool teachers, corporate frontline managers, therapists, computer techs, and police administrators. Two out of three of its students attended its night-school program, which was run out of a former tannery in Manchester. In the 2010s, it made a bid to modernize by updating its offerings and adopting a new name, Mount Washington.

But as New Hampshirites turned their backs on higher education, the same forces that sank Dolly Markey's humble alma mater, McIntosh

College, doomed Mount Washington—its student body evaporated from 4,500 in 2009 to 500 in 2015, its last year in operation.

But before it closed, it issued a criminal-justice degree to the Irish cop who gave Andy a warning for speeding a few weeks before the Button Factory investigation. His name was Michael Maloney, and though Andy hadn't realized it, he was actually more than a cop. He was Greenland's police chief. Antje knew him, because they jointly served on the board of an anti-crime civic workgroup called Seacoast Crime Stoppers.

Chief Maloney wanted to prove that an institutional law-enforcement agency could capture the public trust by overcoming the cold impersonality of the collective and operating with a sense of humanity and humility.

As police chief, he worked to gain the public trust the old-fashioned way—face-to-face contact, on the streets of his small town. No job was too small, so long as it put him elbow to elbow with Greenland's 3,600 citizens (not counting ghosts or aliens). In the wee hours of the night, he showed up in person to reassure homeowners whose alarms had been set off. He ticketed speeders like Andy. He'd once revived a cafe owner who had passed out from heatstroke, and carried fire hoses to a burning mobile home alongside a short-staffed fire department.

Maloney's approach worked. Greenland residents got to know Maloney as a gentle bear—he wore shorts and Hawaiian shirts year-round, but his handshake was iron, cloaked in a bit of pudge. In the tradition of granite-blooded New Hampshirites, Maloney liked his coffee cheap and black. He played the lottery. He fished well, and golfed badly. And he treated the Greenland Police Department like a second family, organizing departmental fishing and ski trips.

Unfortunately for Maloney, the functional and human-centered brand of institutionalism he embodied seemed to be rapidly leaving New Hampshire.

With 1.4 million residents, New Hampshire is among the smallest states by population in the country; more people live in the Bronx. Here, every institution is critical. After losing the McIntosh and Mount Washington colleges, in April 2012 arts-focused Chester College announced that enrollment declines had created a $750,000 deficit; in May, it closed forever. During the same period, St. John International University,

Lebanon College, and Daniel Webster College were freezing salaries and reconsidering existing programming in order to compete for a shrinking pool of students. Within five years, all three would be gone.

And the budget cuts that the legislature imposed on the public college system, which includes UNH, had disastrous consequences. UNH laid off 20 employees, and convinced 110 more to retire early. It froze salaries, scaled back benefits, and hiked tuition rates, which of course made it less accessible to those most in need of its services. Also impacted by the system's budget cuts, New Hampshire's PBS affiliate, which had one million viewers, eliminated twenty full-time positions, cut wages by as much as 10 percent, and gutted the retirement plan. And the UNH Cooperative Extension laid off two dozen employees, trimmed parenting classes, and eliminated programs on money management and food preparation.

I said before that it's difficult for institutions to earn the public's trust, because trust is lost quickly, and won slowly. The truth is actually even worse, according to University of Chicago scholar Tom W. Smith, who found in his research evidence for "a general, master confidence trend." Even accounting for the impact of such individual negative episodes, "there may be generalized forces weakening confidence across most institutions."

New Hampshire was firmly in the grips of a waning confidence trend, and one of the little-recognized forces was the rise of paranormal beliefs, as fueled by the internet, and by a capitalistic ecosystem in which both money and self-identity were becoming invested in the advancement of those paranormal beliefs.

It was a zero-sum game. In order to believe that ghosts, aliens, and cryptids are real, one must believe that the lack of acknowledgment by science, government, and other institutions represents a massive failure of competence or corruption.

Ever so slowly, this dynamic creeps up until one day we wake up in a world that is unrecognizable. It's like they say: if you put a frog in cool water and slowly bring it up to boiling, well then, you're a terrible person. The impacts ebb and flow across the globe, and in March 2012 alone, news reports of emerging outliers offered a vision of what we might expect in a post-institutional future.

In Australia, a medium told a man that he was due an "abundance of riches" as an inheritance, after which he went home and strangled his parents. In Arizona, a gimmick-minded sixty-seven-year-old reverend employed heavily done-up teenage girls to perform demon exorcisms, gaining uncritical coverage from a host of institutional media sources. In Whitby, England, recently elected town councilor Simon Parkes told the press he considered his real mother to be a nine-foot-tall green alien who had appeared to him at various times of his childhood and adulthood. Another alien, which he referred to as the Cat Queen, beamed Councilor Parkes up into space four times a year using a technology that he admittedly did not understand, for the purpose of having sex. He and the Cat Queen had fathered a hybrid alien child named Zarka.

"My wife found out about it and was very unhappy, clearly," he told reporters. "That caused a few problems, but [the sex] is not on a human level, so I don't see it as wrong."

And in France, a village of 194 residents at the foot of the Pic de Bugarach mountain was overwhelmed by twenty thousand white-robed visitors who anticipated that, when the world ended (because: the debunked Mayan calendar thing), aliens hidden inside the mountain would evacuate them to interstellar safety. The visitors lay around, chanted, and prayed in the nude. When one attempted to commit hara-kiri with a samurai sword, the mayor formally asked the French Army to seal the region against UFO enthusiasts.

Right now, these events stand out because they are unusual; but as institutions continue to hemorrhage public trust, these paranormal framings will become ever more common; this paradigm change is what Beau means when she tells her students that a world-changing spiritual shift is coming.

One person who was doing his job in such a way as to forestall that eventuality was Greenland police chief Michael Maloney, who successfully blended his humanity into his role in public service by performing such menial tasks as sitting on a crime watch with Antje, or issuing a traffic warning to Andy.

But Andy would never again see Maloney. Alive.

A Crane, a Lemur, an Ocelot

The month after twenty thousand UFO enthusiasts stormed a French village, in April 2012, Val finally got a chance to shine on her own. With the Button Factory investigation behind her, Beau was off on a two-month vacation, which made Val KRI's senior medium. Andy encouraged her to step up into new duties, including running some of Beau's programs. One of Beau's signature events was a weekly Spiritual Community Gathering, where attendees shared their experiences with the spiritual world and asked questions of a facilitator.

During the gathering, Beau always read a note that had been dictated to her, for this purpose, from a spirit that Beau called "Bob." The notes sounded like someone had flung some fortune cookies, horoscopes, and haikus into a blender and turned the setting to obfuscate. Their intent was not to dictate a literal fact, said Beau, but to bring out one's own internal wisdom.

Though Val led the session, the highlight remained the missive from Bob, which Beau sent to KRI for reading and consumption by the Spiritual Community Gathering attendees.

On Thursday, April 12, Val and her nine-year-old daughter, Juliana, arrived at KRI before six, to ensure that Bob's message was in place and to prep for the event. They made small talk with Shannon and Andy.

Just then, a bit more than a mile away, Chief Maloney was sitting in his cruiser on Post Road, just out of sight of a forties-era white clapboard house with a pitched roof. He had reddish receding hair, and a bristly mustache at the balancing point of his ovoid head. After a career of service in law-enforcement institutions, his love of his recliner, vodka, the Patriots, and his cats were finally beginning to outweigh his love of the job. He was due to retire in eight days.

Maloney's last piece of unfinished business was inside the clapboard house.

Post Road's newest resident was a volunteer firefighter named Cullen Mutrie, twenty-nine, who had a bodybuilder's physique and an air of trouble. Mutrie was six feet, three inches tall, and 275 pounds. He had hooded eyes, and his ears lay flat against his shaved head. Ever since he had moved in six years ago, neighbors said, carloads of shady characters would show up at all hours. There would be loud arguments on the porch in the middle of the night; sometimes they saw him shooting woodchucks in the backyard.

And Mutrie had a rap sheet. At a house party in Hampton, he'd fractured the skull of a man who'd recently begun dating Mutrie's ex-girlfriend; he'd punched a barroom bouncer at the Portsmouth Gas Light; and multiple girlfriends had gone to court to seek protection from his jealous rages, which they said included him pulling their hair, choking them, grabbing them, and pushing them down.

The neighbor complaints about Mutrie frustrated Maloney.

"I can't tell them anything," he would say. All he could offer were vague platitudes. "It's being handled," or "It's being taken care of," or "Don't worry."

The truth was, Maloney had partnered with the state Seacoast Drug Task Force in a long-running covert operation that had built an airtight drug case against Mutrie. The officers were certain that Mutrie was sitting on a pile of narcotics that swelled and shrank with every visitor. Today was the day Maloney would see Mutrie behind bars.

An unmarked van held six DTF members, including team leader Detective Scott Kukesh, a ten-year veteran of the Newmarket police department who had, that morning, provided a security detail for a Seacoast visit from Vice President Joe Biden.

In theory, Mutrie would be unarmed. A recent domestic-disturbance complaint had given officers cause to remove all the guns they could find in the home. Which was a lot. It was like the Twelve Days of Christmas, but with firearms—he had six boxes of bullets, five types of rifles, four 12-gauge shotguns, three Browning pistols, two .22 handguns, and an item that your true love has almost certainly not given to thee, an automatic submachine gun.

Had they confiscated them all? It was hard to say.

A little while ago, one of the task-force members had driven by in an unmarked sedan to scope the house. The overgrown circular driveway held Mutrie's BMW and a car belonging to his on-again, off-again girlfriend, a hairdresser named Brittany Tibbetts, for whom the officers also had a drug-related warrant.

Maloney knew he couldn't be seen until Mutrie was in cuffs. An informant in the drug case said Mutrie frequently complained that Maloney was "out to get him." The two had beefed, briefly, over whether Mutrie could drive across the path of a local footrace on Route 151.

As they considered whether Mutrie was inside, the shadows were getting long, making visibility an issue on the small, high-railinged front porch. Within one of those shadows, a small security camera was mounted, directed at the front door.

Antje stopped by the KRI Center to say hello. She was, as usual, unfailingly pleasant to Val, but an invisible barrier remained between them.

Andy, who had signed off on Antje's KRI membership on Val's behalf, was incensed.

"Antje really betrayed the trust," he said. "She just literally stopped talking to her, treated her like she was shit," he said.

Val hadn't felt comfortable confronting Antje over what had changed. Instead, she turned to Andy, who told her what he had gleaned through Beau. They didn't like hanging out with Val, he said, because her personality was too negative.

Looking back, Antje blames herself for distancing from Val. Instead of accepting Val's experience, Antje said, she focused on how uncomfortable she herself was feeling. "I didn't hold her in high enough regard, to be honest," she told me. "I think maybe I was a little selfish in that."

Before the Spiritual Community Gathering began, Antje said she had to go home. Once she left, Val continued setting up the event. The emotional pain wasn't sharp—in a way, the ache was a soothing confirmation that she lived in an inherently unreliable world in which friends, particularly women, could not be trusted. Beau was teaching her to expect more for herself. And so she would.

Before Val could lose herself in the work, her thoughts were interrupted by a piercing siren. A police car whizzed by on Route 33 at top speed. Soon another screamed by, and more, along with fire trucks and ambulances. It was like a civic parade, run at 100x.

Something had gone wrong at the Mutrie bust.

After a brief powwow, the task-force members decided they would enter the house by force. They pulled the unmarked van into the driveway next to the Greenland Police Department cruiser. Two of Maloney's local officers circled around to the rear of the property, and covered the exit there.

Six DTF members hustled up the porch with a battering ram, wearing raid jackets and bullet-resistant vests under balaclava-style masks. Standing in single file (a tactical formation called a stack), Detective Jeremiah Murphy, thirty-four, swung the ram into the stout oaken door. One lock gave, but another, a steel bolt, refused to yield. They used the ram to knock out a living-room window, but the encroaching darkness, and a drawn shade, prevented them from seeing inside.

They returned to the door, and after a more insistent pounding, at 6:22 p.m., the bolt bent enough for the door to finally yield. The officers at the front of the stack charged inside, but almost immediately got tangled up; someone had moved a large air-hockey table to the front of the door as a makeshift barricade; there was also a kitchen island that prevented them from moving to one side. They were hemmed in, and with the sunlight behind them, they were effectively blinded. The only light they saw was the flash of the muzzle from Mutrie's gun—a Ruger five-shot .357 revolver that Tibbetts had purchased at a Manchester gun show.

The bullets were, seemingly, everywhere.

"As soon as that door opened, that pistol was already up in our faces and as soon as that door got completely open, he was already shooting," an officer later said.

Mutrie, who'd been watching the security feed from his laptop, said not a word as he fired from behind the air-hockey table. The officers abruptly reversed course and spilled back out of the house, off the front porch, seeking cover in the barren front yard. Mutrie advanced, firing. Bullets struck arms, abdomens, chests, and throat.

One officer, an Afghanistan veteran, moved to the right of the porch and began angling gunshots through the doorway and the broken window. One of his bullets grazed the forearm of Mutrie, who stepped backwards into the darkness of the house.

Amid the gunfire, Maloney barreled his cruiser forward, maneuvering between Mutrie's house and the officers, some of whom were lying prone.

Four of the six were hit. Murphy and Kukesh had each been shot in the chest. The other officers screamed into their phones, and scrambled for first-aid kits. Maloney got out and grabbed the bleeding Kukesh from the front just as another officer—the war vet—grabbed him from behind. Together, they dragged Kukesh out of the line of fire.

"I can't feel my arms, Dave," Murphy told one of Maloney's officers, Dave LoConte. They would later dig a bullet out of Murphy to make the ballistics case. "I can't feel my legs."

"You're going to be alright," LoConte practically screamed at him. "Jerry, you've got to stay with me. You've got to stay with me. Just look at me. Look at me."

Greenland officer Theodore Hartmann, whose cruiser Andy and Val had just seen blazing to the scene, pulled his car up behind the others. The house had gone silent, but there was no telling when Mutrie might start firing again.

Maloney waved Hartmann's cruiser over, half-carried Kukesh around the side, and loaded him into the back seat. Kukesh's leg was hanging out of Hartmann's cruiser; Hartmann and Maloney shoved at it, mindful that the shooter was still in the house.

Someone else put another wounded officer into Hartmann's cruiser, and he lit out, headed for Portsmouth Regional Hospital.

The remaining officers were regrouping, screaming at neighborhood looky-loos to go inside, and establishing sheltered locations around the house. Maloney crouched behind the front right wheel of his cruiser, staying low so he couldn't be hit.

The silence extended for ten minutes. Maloney raised his head just enough to peek over the car's hood. Two gunshots, the sound of breaking glass. Maloney collapsed, shot in the head.

One of Beau's foundational teachings, something Val would be very likely to mention during the Spiritual Community Gatherings, is that anyone can develop a psychic ear for murmurs from beyond.

"Everybody on the planet has intuition of some sort. Everybody gets gut feelings. Everyone is like, 'I knew it was my Aunt Betsy calling before the phone rang,'" said Val. "It's about learning the different ways it can show up and then learning how to listen to it. And how to trust it."

Even a self-declared science-centric person like Andy. He'd been taking Beau's classes to learn how to tap in. Now, from the main meeting room of the Center, he looked at the doorway.

There was something there; an indistinct humanoid that seemed to be flickering between two similarly shaped forms. They were both familiar to Andy.

"I can't tell," he said, "if it's Brian Dennehy or Drew Carey."

A New England Irish American, and a trucker-turned-actor, Dennehy had starred in 180 movies but was, to Andy, the parochial sheriff who ran afoul of Sylvester Stallone's Rambo character in *First Blood*. Carey was a comedian and television star who had settled into a role hosting *The Price Is Right*.

Andy knew that different spiritually sensitive people experienced phantoms in different ways; Beau saw full-body apparitions; Antje heard voices; Val felt emotions.

"Most of my readings were expressed in movie characters," Andy said. "When I was first delving into it, it was a lot of really fucked up cartoon characters I saw roaming the place."

He understands this to be a representation of the personality being projected by the spirit, translated into a character that had meaning for Andy.

A Crane, a Lemur, an Ocelot

When the sirens went past, "what I saw was more like a gray blob that was projecting the sensation, I'm Brian Dennehy or Drew Carey."

The spirit hovered near the doorway; this made sense, as Beau had established a protected space in the meeting room that did not allow most spirits entry into the room. After a minute or so, the apparition vanished.

The phone rang. Andy picked up. It was Antje.

"Somebody's been killed," she said.

Maloney's body lay on the ground for more than thirty minutes, until an armored BearCat vehicle trundled to his position and confirmed what they all knew. The amiable Greenland police chief was dead, his unfired pistol having fallen from his hand onto the ground beside him. His wife had been organizing a retirement party for him at the Greenland Country Club. One hundred and fifty people had already accepted invitations.

After Mutrie's initial attack on the officers, he and Tibbetts had retreated to the basement. Mutrie used a 9mm purchased by his father in 1989 through a basement window to hit Maloney's barely exposed head, twice. It was such an unlikely shot that the drug task force thought it had been fired by a scoped sniper rifle from on high.

At 2 a.m., a robot went up the porch stairs and began a room-by-room search for Mutrie. It eventually found him with the 9mm handgun in his lap. Tibbetts's legs were crossed, and her head was resting on his thigh. They were surrounded by burnt cigarette butts, twenty-seven grams of cocaine, three vials of steroids, their cellphones, and a bottle of Mountain Dew. Mutrie had used the 9mm to fatally shoot Tibbetts, and then himself.

At KRI, before the details of the tragedy spread, Val held the Spiritual Community Gathering as scheduled. It was smaller than anticipated. There were a couple of no-shows, including a professional marketer and amateur Reiki healer who had RSVPed, but didn't actually show up.

Afterward, she tidied up the space, and was one of the last people to leave. Because Antje had said the shooter was still at large, Andy walked Val and her daughter to her car in the parking lot.

The spirit Andy had seen at the time of Maloney's death was a spiritual rebus to be solved.

"I had no idea it was a cop," said Andy. "I had no idea what was going on in Greenland."

But as Maloney's picture flashed on local television stations, Andy realized that it was the officer who had pulled him over a couple of months ago. When he and Antje spoke on the phone again, she made the connection. Dennehy was a small-town cop. Carey represented a sense of humor.

"Holy shit," said Andy. "That was him."

He later said that he found the night's note from Bob to be timely, poignant, and freaky. Though it was typically shared only with event attendees, he offered it up to the broader community so that they could see the relevance.

"Gratitude is supple and futile if left abandoned," Bob had told Beau. "The step around the corner is the true engine of gratitude. A crane, a lemur, an ocelot are not in picture books. They are seen later in life. If left with only picture books one never expands. It is around the corner, what we really want to see, that is gratitude experienced. I find no purpose in keeping Pandora's box closed."

And when it comes to the consciousness of Chief Michael Maloney, who are we to believe? Did it, as the church teaches, ascend to Heaven? Did it, as science teaches, cease to exist? Or did it, as a group of increasingly prominent players in the local paranormal-enthusiast community teach, make a pit stop outside KRI, and then elicit an inscrutable message from a high-level spirit named Bob?

Nineteen

THE OUIJA BOARD

Mark regained control with the impatiently traced words:
"Every scribe here wants a pencil on earth."
Not until the middle of summer did we achieve that sureness of touch which now enables us to recognize, intuitively, the presence of the one scribe whose thoughts we are eager to transmit. That the story of Jap Herron and the two short stories which preceded it are the actual post-mortem work of Samuel L. Clemens, known to the world as Mark Twain, we do not for one moment doubt.
—Emily Grant Hutchings* (And Mark Twain. Maybe.)

No one believed that Chief Maloney's spirit arrived at KRI's doorstep by chance.

The truth was, after a year of inviting contact with other planes, the Center itself was becoming haunted. Andy called it the lightning rod effect.

"Anything dead within a mile, we're like a beacon," he said. One night, while Val and a friend were watching a movie in the meeting room, there was a loud crash on the roof. They ran outside with Andy, but found

---
* This is from a 1917 novel that author and medium Hutchings claimed was written by Mark Twain, seven years after his death. Hutchings said that the novel was dictated to her and medium Lola Hays from beyond the grave by the deceased Twain through use of a Ouija board. After being sued by Twain's widow, Hutchings destroyed all copies of the book.

nothing. A spirit, they think. Elsewhere in the Center, recorders picked up whispers. Items moved or broke when no one was watching. A lot of the activity seemed to be centered around Mike's desk. Photos on the adjacent wall fell without reason. Electronics were continuously fussy. Another time, a small ceramic raccoon—one of a set of four that Mike brought into his workspace—toppled from his desk's upper shelf.

Mike didn't think too much of the raccoon. But then it happened again. And again. This was a real puzzle. The raccoon was designed to hang over the edge of the shelf, but its center of gravity was well away from the edge. In addition, it was held in place by a strip of adhesive. And yet, sometimes it fell three or four times in a single day.

They thought the raccoon was being moved by a spirit, so Andy set up a camera to monitor Mike's desk. The first day, the camera showed nothing. The next day, nothing. Then another day, another, another. Nothing, nothing, nothing.

Andy kept the camera up for a full month. The raccoon never budged. Whatever it was didn't want to be filmed.

Because of all the activity, Andy sometimes ran small-scale ghost hunts for new and aspiring investigators right at KRI. During one training session, Mike, Andy, Val, and Antje walked a couple of trainees through Andy's rigorous equipment protocols. Mike set up the video camera at his desk again.

"Don't bother," said Andy. But Mike set it up anyway. They paired up, turned off the lights, and began their mock investigation of the Center—into and out of each office, and then out into the retail center, which, despite endless hours of work from Andy, was still nowhere near finished.

The Center was uncharacteristically quiet.

At some point, the investigation wound down. The lights remained off, but the group relaxed into conversation, congregating around a small coffee table and a couple of couches in the back room, near Mike's desk.

They began chatting. One of the great things about KRI was that there were always any number of fascinating topics to discuss. For example, one of Beau's students was honing her skill to become a specialist in talking to animals—alive or dead. She could ferry messages between a dog, say, and its former owner, in the form of feelings, pictures, words, and "knowings." She could also scan a living pet for energy imbalances.

This form of animal communication is far more personal and dramatic than what science acknowledges. One way that we know how to communicate with animals is by analyzing their natural language, something that we understand better thanks to UNH avian communication expert Karina Sanchez, and UNH marine acoustic ecologist Michelle Fournet. Sanchez's work helped demonstrate that bird songs are regional and cultural, while Fournet uses recordings of humpback whale songs to try to define their meanings, and have actual dialogues with living whales.

At some point, the discussion at KRI turned to Ouija boards. Beau's anti-Ouija policy reflected a commonly held belief among mediums that was amplified during the demonology era of Ed and Lorraine Warren. A Ouija board invited communication with malevolent spirits.

"It's like taking off your clothes and running naked down the street saying, 'Pay attention to me,'" said Antje. "You don't know what you're going to get."

There's no definitive list of what sorts of malevolent spirits are hovering invisibly around us, but people have posited a long list that runs the spectrum from powerful demons that can possess a human body to various forms of spiritual parasites. The parasites are thought to be common, and often attach themselves to an unwary ghost hunter. Some feed from their human host's life force, which can result in depression and a loss of energy, while others feed specifically from negative emotions, which gives them an incentive to foster negativity in various ways.

This is why, when Antje talks to other aspiring intuitives, she often gives them a piece of unambiguous advice: "Don't frigging use Ouija boards."

This is also why I was surprised at what the KRI members decided to do after their training session.

"Somehow," said Antje, "we decided to drag out the Ouija board."

Beau wasn't there, and their reasoning leaned on another of her teachings—that a trained intuitive was capable of fending off any attacks from negative entities. Val and Antje were both eager to prove they could protect the group. Mike didn't object. And Andy believed that he was more or less impervious to these sorts of dangers.

And so they lit some tealights, put the board on a low coffee table, and sat on the floor around it, big couches at their backs. After some minutes of failure to attract a spirit, Val noticed Andy oscillating his head back and forth in the candlelight. The movement was unnatural, somehow. Unnerving. The planchette sat on the board as if glued into place.

Andy said he had heard a noise. Something moving.

They all hushed, trying to pick up on whatever it might be. Val felt a prickling sensation on her spine.

"Something was behind me," she said. "My back had some open air. I didn't like it."

The quiet was broken by a loud clattering from right beside the group. Mike jabbed a finger at his desk as Val shrieked and scrambled away, winding up on Antje's lap.

"Raccoon!" shouted Mike.

The first time I visited Andy's apartment, he pulled video of this moment up on his laptop. There they all were, speaking animatedly, when Andy said he thought he'd heard something. The group tensed, and they began talking over each other in urgent whispers. The camera clearly showed the raccoon figurine tumbling of its own accord from the shelf, landing on the desk and then the floor with an audible crash.

"It didn't fly off," said Antje. "It didn't fly across the room. It wasn't that dramatic. But it also didn't roll off."

Was this a Solid Phantom? The whole fate of the world, the whole understanding of the universe, wrapped up in this one little piece of etched ceramic tchotchke, this worn strip of adhesive tape, this naturalistic representation of a furry little hand-washer. An inert bit of matter like this doesn't move on its own; an outside force must act upon it. On that the believers and skeptics agree. But what was the outside force—something mundane? Or something so foreign to our understanding of physics that it might as well be magic?

I began to appreciate something that Andy often said to me. He had amassed few replicable experiments, but piles of what he called evidence. His observations often took unexpected forms, in a unique set of circumstances, in a way that could not be recreated in a laboratory. They were like handicapped field biologists, observing the effects of little-understood,

The Ouija Board

invisible creatures, and trying to infer something meaningful. They were microbiologists without microscopes, astronomers without telescopes, Jane Goodall among ethereal chimpanzees.

Was video of the falling raccoon evidence of a ghost? It wasn't a double-blinded study. But it wasn't nothing, either.

I suggested to Andy that perhaps a constant minor vibration—created by, say, a desktop printer—had slowly edged the raccoon to the very precipice, allowing a simple air current to create a spectacular effect.

This is a fundamental difference in mindset between the skeptic and the believer. I saw the raccoon fly off the shelf, and cast a net for explanations that accommodated all known scientific principles and knowledge. I wanted a mundane explanation, even one that represented an amazing coincidence.

Andy, as a believer, accepted the simplest explanation from a larger universe of possibilities, one that allows for unproven paranormal causes. From his perspective, a ghost was much likelier than my vibration hypothesis. The tape was sticky, he said. It could have vibrated for ten years and the raccoon wouldn't have budged. He handed me the actual raccoon. It was heavier than I expected. I put it down, and saw how it was designed to hug the edge of a shelf with its tail. Even without the tape, it was at least as rooted in place as a stapler, or a tape dispenser. To make it unstable, one would have had to deliberately shift it so the tail was beyond the edge.

I later suggested a different mundane explanation to Val, Antje, and Beau. Could one or more of them have set up a prank by rigging the raccoon to fall? Though I didn't say it, I was thinking of Mike, who had set up the camera at precisely the right time, who owned the desk, and who had amplified the reaction by pointing and shouting.

But everyone said that this would have been unthinkable. Pranks like this simply didn't happen at KRI. If one of them had faked the evidence to fool the others, it would have violated the mutual trust that allowed them all to tiptoe together onto the thin ice underlying the hinterlands of socially acceptable thought.

Twenty

YOU REALIZE LEX LUTHOR LOSES

In the morning, a note from the ghost reminded them that the money was due.
—Gaston Leroux, *The Phantom of the Opera*

There was, of course, a constant need for money. They would have liked to focus on a schedule of pure paranormal research, but every investigation involved travel costs. Every month, a rent payment came due. Every piece of equipment had a price tag. Their effort could only be sustainable if it was financially sustainable.

KRI sailed along on the optimism of the small-scale entrepreneur: if they continued laboring uphill, they would at some point heave themselves over a mountaintop, and rollick their way down a grassy slope, into the gilded curbs of easy street. One obstacle was that, a year after they identified their three-legged revenue plan, the third leg remained purely aspirational.

Event fees and donations were flowing in. So were monthly sublets, including from Beau, who paid $500 a month for her office. But the third leg, the retail center, had not materialized. Andy was still renovating the retail space.

"I love hands-on crap," he said. "I could do construction for the rest of my life. I enjoy making things."

He approached each facet of the project with a craftsman's reverence. The walls, he decided, should be entirely rebuilt. He was also hand-crafting a beautiful retail counter. Andy's approach to the work was largely intuitive. He didn't sketch a blueprint, or look up instructions for similar projects on YouTube. Instead, he measured the space, and let the work flow from his hands, barely thinking about it. It was a process that he loved. But Andy's labor of love was, to the others, labor lost. The retail center was sucking up all of Andy's time, and he had nothing to show for it.

Antje saw it as an organization problem.

"All he did was do carpentry work," she said. Andy could be downright scatterbrained about other duties at the Center. He would drop balls, but was reluctant to delegate decision-making to the others.

"Andy is probably one of my best friends," said Val. "But he didn't do the majority of the work at the Center. He doesn't know it's daylight savings time without someone constantly there to remind him."

From Andy's perspective, the primary problem was a lack of resources. They couldn't mount an effective advertising campaign. They couldn't staff a front desk. They couldn't pay for a web-design team, or a construction crew to come in and fix everything. All he had was volunteers. He felt like it all fell on his shoulders.

But everyone agreed that KRI was having money problems. And money problems caused integrity problems.

One of the founding principles of KRI was an adherence to the truth; but that principle was threatened every time there was a chance to make money from something other than the truth. The Social Saucers events were just one example.

"The more you meet people, the more you understand how nutty they are," Andy said. "And how absurd their reads on things are." The scientific standard that Andy was so passionate about was not what attracted the gate fees and donations they were receiving. In fact, KRI donors often seemed to be most interested in the least credible claims.

"There's a problem," Andy realized. "Science makes the truly strange mundane. There is no interest in the mundane."

Because of this, KRI largely abandoned its assurances to the public that it would only platform credible ideas. They hosted Alchemist circles, a metaphysical hypnotist from Dover who taught astral projection, and a series from "The New Paradigm Multi-Dimensional Transformation Organization," better known under its former name, Shamballa Multi-Dimensional Healing. Rather than screen questionable groups out, Andy expressed his standard largely by arguing with presenters after their presentations. Or, if an event attendee stated an erroneous position during a group discussion, Andy would correct them; if he didn't like them, he corrected them aggressively.

"He would tell people why they were wrong," said Val. "It wasn't 'I disagree.' It was 'You're wrong and I'm going to tell you why you're wrong.' I tried so many times to tell him how he could say that in a nicer way. But he doesn't listen."

Antje put it differently.

"He, uh, he will rip people to shreds. To shreds," said Antje. "Yeah. For not—you know, for violating his perception of qualification and, and competency."

I spoke to Andy about being KRI's barometer of how real something was. He said that, while he agreed people are entitled to their viewpoints, in practice, this had to be balanced with reality.

"I don't want to break them," Andy said. "But it's inevitable. Some of the things people say are silly."

When Andy told me that some KRI presenters had silly beliefs, I wanted to learn more. I knew Andy believed in spirits, aliens, the ability of people to bend spoons with their minds, the possibility of a hollow Earth, and spiritual possession. Where was the line at which the beliefs became so bizarre that he didn't accept them? I tossed something out to get the conversation started. Was it mermaids?

He hesitated.

"I actually believe in the reports of mermaids," he said. "Somewhat."

I tried again. Surely, I said, he didn't believe in Atlantis.

Again, Andy seemed torn. He paused, then spoke slowly.

"I don't want to sacrifice my credibility," he said. But, he told me, the idea that there was once an ahead-of-its-time civilization called Atlantis was not as far-fetched as it might seem.

"Okay," I said. "So what are the beliefs that are so out there that you—who believes the category of stuff that you believe in—say it seems outlandish and improbable?"

Andy rose to the challenge, and referenced a vibrant conspiracy theory about a covert, but widespread, planetary takeover by an advanced race of lizard people.

"Outlandish and improbable," he said, "is lizards posing as humans so well that most people don't know, but only a special few can see it when their eyes turn lizard-like for a second and go back. And a LOT of people believe in lizards."

He clarified that advanced lizard civilizations were not the objectionable element of this belief. Velociraptors could have easily, he said, over seventy million years, evolved into intelligent quasi-human beings with advanced technologies and then decided to leave the planet. It was the idea that some people could detect lizards among us by noticing occasional reptilian eye-flickers that he found to be outlandish.

There was another way that Andy expressed his adherence to standards—though KRI rejected few, if any, speakers on the basis of subject matter, Andy took to verifying the academic credentials of presenters. This approach was endorsed by The Reverend Beau.

"I'm a Reverend," Beau told me. "'Cause I was like, 'Is this for real? Can you really do this?' So I went online and did it. So I, I'm a Reverend by joke."

Beau said she was taken aback by the ease with which some credentials can be attained.

"When they said, 'Yeah, thanks for your $15, you're a Reverend.' I was like, wait a minute. And then, for $35 or $45, you could be a doctor."

Credentialing came up when world-renowned spiritualist Dr. Steven Farmer agreed to visit KRI during a trip to the East Coast.

Everyone knew Dr. Farmer's work. He'd written the definitive guide to animal spirits. He believed that, through meditation, one could establish a connection with a spirit that drew its power from its wild and instinctual nature. This animal conferred its primordial character traits onto the individual. Confusingly, he says animals with no physical representation in the material world, such as dragons or unicorns, can also be spirit animals.

Andy was a fan. For years, he had been acquainted with his totem animal, a spirit jaguar whom Andy named, he gleefully revealed, "Mick Jaguar." Beau once held a session in which she asked people to commune with each other's spirit animal, and Andy's jaguar had shown up in the consciousness of his meditation partner. She told Andy his spirit was actually an ocelot, a feline with jaguar-like markings that is only slightly bigger than a house cat.

No, said Andy. It's not an ocelot. But you're close.

She went back into meditation, and when she came out, she had a message from "Mick Jaguar," who was, she affirmed, an ocelot.

"Andy likes to think that everything about him is bigger than it is," the spirit animal had said.

Andy roared with laughter. "I call that a hit!"

But Farmer's teachings on spirit animals were more impressive than Farmer's credentials. His PhD came from Madison University, an unaccredited diploma mill that offers degrees for a flat fee.

Once again, the goals of scientific rigor and inclusivity were at odds. Andy decided that Farmer could only come if he dropped the "Dr." from his name in promotional materials for the event. As far as he was concerned, Farmer's credentials were not legit, and he pulled no punches.

"You can come in, but not calling yourself doctor," Andy said. "I check. If you have a quack doctorship, you can't call yourself doctor." Andy refused to back down.

The talks with Farmer quickly fell apart, to the aggravation of KRI's other members. Val felt that having a big name like Farmer was worth a bit of fudging.

"Is it fair to have this piece weigh so heavily when we have all this on the spiritual side?" she asked. "Maybe he doesn't have the legitimate credentials but we have book sales and countless people wanting to speak with him."

Farmer was just one of several potentially lucrative speakers who had more profile than credentialing integrity.

"Things like that happened very regularly," said Val, "and it was very frustrating."

As KRI lurched along, Antje was working on a science project that had a potential financial upside. She wanted KRI to offer a unique spiritual experience to the public. She wanted to create a psychomanteum.

A psychomanteum is an enclosed space that screens out sensory clutter and compels its occupant to focus on a mirror, lit only by a faint candle. It springboards from an ancient practice of scrying with a looking glass, reflective pool, or crystal ball.

There is scant research on them, but one 2002 study by Dr. Arthur Hastings, a founding faculty member of the paranormal-themed Institute of Transpersonal Psychology, found that during a single guided psychomanteum session, roughly half of bereaved users experienced a range of contacts with the spirits of their loved ones; these included not only images of the spirits, but voices, smells, touches, and conversations. Both this study and another by Hastings showed that these experiences tended to alleviate symptoms of grief—bereavement, guilt, sadness, and fear—to a much greater extent than visits to a medium. One research paper suggested that this is because a facilitator at a psychomanteum is focused on the needs of the living, while a medium is often more focused on the needs of the dead.

KRI, Antje realized, could have the only psychomanteum within a thousand miles, and perhaps the only one in the country open to the public. People could sign up for sessions, and KRI could also advance its scientific mission by conducting unique research.

To learn more, Antje called the country's foremost expert on psychomanteum construction: a man named Raymond Moody, best known for popularizing the idea of universal near-death experiences, such as floating above one's body, or traveling toward a light in a dark tunnel. Moody gave Antje an exhaustive list of specs. The space had to be soundproofed, and equipped with a zero-gravity chair. The chair and mirror had to be at a particular angle from the ground, and from one another.

When Antje hung up, her mind was already chewing on details. There was a small unused space beneath the stairs between the two levels of the Center. Here, she thought, was the perfect place to bring people face-to-face with their departed.

In the culture of fiscal scarcity at KRI, one bright spot was the Psychic Sampler events, which they offered twice a month, on weekends. Psychic Samplers operated on a bordello-inspired business model, and like a bordello, they always brought in money. A customer would consider a lineup of mediums sitting at small tables in the meeting room. For ten dollars, they got fifteen minutes in a private session with the medium of their choice.

The money was paid to the coordinator, typically Val, though she sometimes filled in to do readings herself when they didn't have at least five mediums on offer. Antje also did sampler readings sometimes. Once she told a client that her grandfather's spirit was there, and wanted to dictate a pickle recipe.

"And she turned absolutely white," Antje later said, recalling the client's face. "She said, 'We have been looking for that recipe for years.'"

The new mediums were mostly culled from the students at Beau's intuitive development classes. For them, the chance to fine-tune their skills in a professional setting was a boon. Andy sat with each new reader to evaluate them.

Beau's students were mostly, but not entirely, women, which makes sense, given that the broader spiritualist movement is dominated by women. Beau observed that, generally speaking, women are more flexible in their spiritual journeys. A balanced seeker assesses each experience with something closer to a blank slate, one that is open to multiple interpretations.

Men were inclined to be the extremists of the psychic world.

"They tend to just need titles and definitions and a place to land," said Beau. "So when you talk to a man, they last night dealt with the

Succubus, and today they're on the seventh level of the fourteenth dimension of Uranus."

Once a man finds that sort of lens, Beau said, he stalls on his journey, trying to fit the whole world into that paradigm.

"They get very stuck on, well, how does that relate to my starred seed existence from my fourth AI generation of the Seventh Alien tribe?" she said. "And it's like, okay. Um, there's more than that out there."

One example of a man who had landed, and landed hard, met Andy to audition as a new reader at the Psychic Sampler. He was a friend of one of KRI's regular event attendees, Roberta Gerkin, forty-seven. ("She was very tall. She might have had acceptance issues," Andy noted sagely. "She was probably close to six feet.")

Gerkin had a strong chin that seemed to weight the corners of her mouth, and hooded, thoughtful eyes.

Her friend wore black pants and a black magician's vest over a white button-up dress shirt and a necktie. She had met him during last year's Halloween event at the Governor's Inn in Rochester. As she introduced him, the man interrupted.

"I'm Lex," he said. His voice was high-pitched and a bit raspy. They understood that he didn't want Gerkin to say his actual name.

"Like Lex Luthor?" asked Andy.

"Yeah," said Lex.

"I'm like, 'Wow, you realize Lex Luthor loses in every single issue of Superman,'" Andy said.

He thinks he's way too smart, Andy decided.

Andy didn't like Lex, but held his tongue as Gerkin waxed enthusiastically about his abilities.

Lex had adopted the vest as a signature look for its slimming properties, after donning one at a community theater performance. It gave him the dapper look of a stage magician, or a bartender laboring under a dress code. He wrote horror and fantasy scripts, and aspired to a horror stage production. A 2006 graduate of UNH with a degree in theater, he was also a fourth-degree black belt in karate who taught at the dojo he had started attending as a child in Kittery, Maine.

You Realize Lex Luthor Loses

When Andy saw what tarot card deck Lex was using, Andy practically rolled his eyes. "He had this very pretentious deck. The Aleister Crowley magician deck. Which is, every card looks crazy evil."

Lex did a reading for Andy.

Sometimes, readings stuck out to Andy, either for being really good, or really bad. This was neither.

"He was meh," Andy determined. "Could have made it up, could have not." He agreed that Lex could sit and participate as a reader in the sampler.

What Andy didn't know was that Lex's beliefs had landed, hard, somewhere among the pages of the fantasy novels he enjoyed. He believed that there was a distant dark realm, beyond "The Veil of Separation," where a superior class of beings interacted with one another, and carried on their own agendas. These beings held influence on Earth by manipulating humans. Four or more otherworldly personas sometimes asserted control of Lex's being.

Gerkin, who was older and more experienced than Lex, was a bit of a mentor for him, at least when it came to tarot. She did individual readings for his various personas, including Cyrus, Wild Card, the Nameless One, and a fierce paladin named Dark Heart. (He also used these characters in the fantasy roleplaying game Dungeons & Dragons.)

Once, when Gerkin did a reading for Dark Heart, Lex told her that Dark Heart had been driving him to feel extreme impulses for sex and violence. It was a really difficult time, Lex said, because his girlfriend, Kat McDonough (she was eighteen, he was twenty-eight), was away at summer camp for two weeks. Without her to have daily sex with, he said, Dark Heart's violent nature was becoming impossible to contain. To appease Dark Heart, Kat had arranged through an online message board for a surrogate to come and have sex with Lex for the two-week period. But the surrogate had not shown. Lex cut to the chase. Would Gerkin give Lex a blow job to prevent Dark Heart from becoming murderous?

Gerkin was recently divorced. Later, she would say that, at that period, "sex was nothing" to her. And so, she successfully averted the crisis, via fellatio.

Not even Gerkin knew about Lex's unusual bucket list. He planned to get into law enforcement, and then join a SWAT team. He would use that experience as a springboard for a career as a private detective, and then somehow assemble an army of "minions" to raze the world on behalf of outsiders everywhere.

He worked at a Best Buy.

After the sampler ended, Andy remained unimpressed.

"I think he was expecting to be the hero there," he said. "That everyone would be, 'Wow, that's amazing.' And like, 'Dude, you're not even good.' No one was impressed."

When Lex left with Gerkin, he seemed disappointed that he had not been the center of attention. Val was happy to see him go. He had made her a bit uncomfortable.

He was, to her, "a little weird. But not murderer weird."

But the man who wanted to be called Lex was more than a little weird. As it turned out, he was murderer weird.

Twenty-One

FOR THAT I AM TRULY SORRY

Alice could not help her lips curling up into a smile as she began: "Do you know, I always thought Unicorns were fabulous monsters, too! I never saw one alive before!"

"Well, now that we have seen each other," said the Unicorn, "if you'll believe in me, I'll believe in you. Is that a bargain?"

"Yes, if you like," said Alice.

—Lewis Carroll, *Through the Looking-Glass*

About a month after meeting Lex, on Saturday, October 13, 2012, Val and Andy talked on the phone about a piece of breaking news on local television. A local man named Seth Mazzaglia had been charged with second-degree homicide in the death of a missing UNH student.

The report kept flashing pictures of Mazzaglia, who had thick brown hair that extended down his chops to wrap his chin in a little fuzzy blanket. He wore a slimming, shiny vest.

"Holy shit," said Andy. "That's Lex!"

Shortly after, Val sent out an email to the group.

"It is now confirmed by a photo in the Portsmouth Herald," she wrote. "The man who is being charged with killing the girl from UNH came to the Center as a friend of Roberta's and did readings at a Psychic Sampler, asking to be called 'Lex.'"

They didn't know what to make of it. Val sent Gerkin a note of support from the KRI team. Though Andy had been unimpressed with Lex as a tarot-card reader, he was a long way from believing he was a murderer.

"I met the guy and there was *nothing* scary about him," Andy wrote in an email to Val, Mike, Antje, and Beau. But as a murder trial began to unfold, complete with testimony from Gerkin, it became clear that she had befriended a monster.

"He was a serial killer," Andy eventually concluded. "He just got caught on his first murder."

When Seth Mazzaglia met Kat McDonough, she was a seventeen-year-old theater geek in Portsmouth, often helping her mom in her pottery shop after school. Within months, Mazzaglia had gone from bonding with her over a shared interest in LARPing, sci-fi, and martial arts to convincing her that she, too, was under the influence of other entities from beyond the Veil: Skarlet, Anay, Violet, and Kitty. Once Mazzaglia and McDonough declared their love for one another, he convinced her to move into his apartment, and soon after that, she broke off all contact with her family.

When McDonough returned to the apartment after the summer-camp trip during which Mazzaglia sought oral sex from Gerkin, he plunged McDonough into a nightmare. He said that, because she had broken a promise to provide him with a sexual partner, she would now submit to sodomy on command, until such time as she found him another woman to have sex with. Under intense pressure from Mazzaglia and the Dark Heart entity they both believed was controlling him, she asked a pretty young coworker from Target—UNH student Lizzi Marriott—to come hang out with her and Seth to watch movies.

Later that night, at about 10:50 p.m., Kat called Gerkin and asked her to come over right away. About twenty minutes later, Gerkin and her boyfriend, Paul Hickok, walked into Mazzaglia's apartment kitchen. Every available surface was a jumble: boxes of Ritz crackers, quick-cook oats, gallon jugs of water, salsa, candles, and paper towels. In the living room, the beige carpet was cluttered with dirty laundry, video games, electronics, books, CDs, movies, and Lizzi Marriott's dead body.

She wore only underpants, and two plastic grocery bags over her head. Mazzaglia told Gerkin that Lizzi's death was due to an extreme BDSM session that had gone too far. He'd blacked out, he said, and awoke to find her dead.

At Gerkin's urging, they removed the bags, but it was clear that the time for emergency resuscitation had long since passed. Gerkin urged him to call 911. By the time she left the apartment, Mazzaglia indicated that he would.

"I was an optimist," Gerkin later said. "I like to believe, but not anymore, that people will do the right thing if given an opportunity."

But a few days later, police knocked on Gerkin's door. Mazzaglia had been arrested. Gerkin agreed to wear a wire and began a series of conversations with Kat, delicately trying to get her to admit that Lizzi's death was not an accident.

While in custody, Mazzaglia orchestrated a series of improbable plans to evade the charges. He sent McDonough letters describing their cover story, which involved McDonough sitting on a restrained Lizzi's face until she asphyxiated (he drew accompanying diagrams). He plotted to kill Gerkin and her boyfriend, as witnesses. He even tried to convince a drug-addict cellmate to take $1,000 out of Mazzaglia's bank account, buy drugs, and sell them to raise $15,000, and then use that money to fund an elaborate escape attempt that included weapons and a boat to help him get overseas. The drug addict took the $1,000, spent it on drugs, and told officials about the plot.

In the end, the truth came out during a lengthy trial.

While they all watched a movie together, McDonough and Mazzaglia had suggested a game of strip poker, to which Lizzi had agreed. Then they suggested sex, to which Lizzi had not agreed. When she returned her attention to the movie, Mazzaglia put on a pair of black gloves, placed a rope around her neck from behind, strangled her to death, and then raped her corpse. McDonough testified that she and Mazzaglia dumped Lizzi's tarp-wrapped body at the mouth of the Piscataqua River in Portsmouth Harbor.

Even when Mazzaglia had a moment in court to mitigate his monstrous image, all he offered was a blame-shifting nonapology that sounded like a sociopathic corporate PR hack run amok.

"I did not rape and murder Elizabeth Marriott. However I do understand the Marriott family's pain and I did play a part in covering up her death, a mistake that I tried to correct when investigators came to me. I showed them exactly where I had left Lizzi's body. Unfortunately they were unable to recover her and for that I am truly sorry."

He was sentenced to life in prison without parole. For her cooperation, McDonough was given a deal that suspended most of her sentence and required her to serve only three years in prison.

The facts of the Mazzaglia case were defined by institutions. But its meaning and context were up for grabs.

An amateur astrologist named Anthony Wynands made much of the birthdates of Mazzaglia and McDonough—he blamed the star Spica, located in the last ten degrees of the tropical zodiac of Libra, which he said created a "spooky influence."

"Mazzaglia had his Saturn-Pluto conjunction conjunct Spica; McDonough had her Venus-Jupiter conjunction conjunct Spica," wrote Wynands.

Wynands noted that Spica (actually a double star 250 light-years away) represented an ear of wheat in the left hand of the virgin in the constellation Virgo—and fermented barley wheat, he went on, with heavy import, was a key ingredient in the consciousness-expanding drink of Greek mythology, kykeon.

This is the sort of explanation that is, for those who believe in such things, beguiling, and is also, for those who don't, a cause for concern. It absolves McDonough and Mazzaglia of their personal responsibility, and also, more importantly, our collective responsibility to build a world in which such murders are less likely to happen.

Part of Mazzaglia's social support network came from Mazzaglia's time as a student at UNH, where he spent time hitting on the members of a pagan circle, a group of students at UNH who celebrated such concepts as fairies, pixies, and Wiccan powers.

Mazzaglia's dream of world domination was one of those smoldering ideas waiting for the right circumstances to burst into flame. Unfortunately for him, he existed in an America in which there was just enough of a spiritual community to give him encouragement to act rashly, but also enough dominance by institutions to bring him to justice relatively early in his planned career of brutal rape and murder. But what if we had adjusted our paranormal-to-institutional ratio? If there was no Wiccan group at UNH, and no tarot-card scene, and no encouragement in the form of someone willing to give him a blow job as a spiritual solid, then perhaps his fantasies would have remained in the realms of Dungeons & Dragons. But if, instead, those elements were stronger components of the zeitgeist, and institutions were comparatively weaker, then perhaps his blood-soaked dream could have come closer to reality.

But UNH, at least, did not understand the Mazzaglia murder in those terms. In 2011, the campus signed off on a UNH Paranormal Club, which quickly grew to thirty members and undertook ghost hunts in the campus's many historic buildings. Both the UNH marketing team and the student paper wrote lighthearted features about the group's findings of mysteriously moving elevators, and a female ghost that roamed the wooden floors of Smith Hall. This is, on its surface, harmless, until the tide of paranormal enthusiasm fuels a monster like Seth Mazzaglia.

It's clear that generalized trends are tilting America toward a new age of New Age thinking. As institutions decline, it's useful to think about what will remain, and what will not. Will the average American be more like Seth Mazzaglia, who centered his entire life upon an individualistic view of the universe? Or will this be a change of window dressing, rather than substance? In other words, are we simply swapping out the hospital chaplaincy for a medium? Or are we also swapping out all the doctors for Reiki healers?

The Ghost Lab

For that answer, we turn once again to UNH. In the late 1960s, when social investment in colleges was universally embraced, UNH underwent a tremendous expansion: McConnell Hall was built to accommodate the business and economics programs, while the Horton Social Science Center was built to house the sociology and anthropology departments. In the mid-eighties, there was also an extensive remodel of Conant Hall (one of the original five campus buildings, circa 1893), to better house the psychology department.

But by the 2010s, programs had been reprioritized to reflect the school's new financial reality. When McConnell Hall grew old and tired, the business and economics programs were moved into a gorgeous new building of blondwood and glass. Meanwhile, the venerable, but not venerated, sociology and psychology departments were packed into McConnell Hall, which campus officials referred to as "sturdy but worn."

And it was in McConnell Hall that the answers lay, in the mouths of those with creaky knees sitting in creaky chairs in a handful of cramped, concrete-walled offices.

This was the workplace of Victor Benassi, the seventy-year-old psychology professor whose many pupils included a bright master's student named Andy Kitt. Benassi was the coauthor of a wide-ranging paper in which he took a sociological approach to analyzing various expressions of belief in the paranormal.

Benassi pointed to Boyoway (now Kiriwina), a large coral island off the coast of New Guinea. In a centuries-long tradition that extended well into the 1900s, the men in the villages along the western coast would catch fish to feed their community, and to trade for yams with inland villagers.

The western edge of the island was a large, sheltered lagoon, and fishing the reefs there was a fairly simple affair for the local villagers. Fishermen pounded the root of a creeper plant called *tuva* into a sticky mess, and wrapped it in a package of leaves. Then they paddled out in a canoe and spread a net over a patch of reef. Using a pointed stick, they lowered the *tuva* package into a hole in the coral reef. In the calm waters, the root's secretions—a deadly poison—accumulated in the hollow until dead or stunned or fleeing fish blundered into the waiting net. It was safe, and it was reliable. It was like poisoning fish in a barrel.

The other primary fishing method targeted sharks in the open sea. Each season, as they began to build fishing canoes, a resident magician was given gifts, and cast a spell triggering a series of strict taboos for the village. No sex, no hammering of wood against wood, no noisy games. No bodily adornments, no hair-combing, no self-anointing of women with coconut oil, no grass petticoats, no strangers. After weeks of these abstentions, the magician rubbed herbs over the boats and cast charms over them.

When the men were ready to take their spiritually charged boats into the water, the magician sat on the platform of his house with his legs wide open, and his penis exposed, singing a song to the sharks.

Out of sight of land, the tiny canoes were subject to dangerous swells. The fishermen would lower a bent stick threaded with coconut shell segments into the water and shake it vigorously. The clacking of the shells mimicked the sound of a shoal of fish jumping out of the water. The shark would rush in to bite at a large, baited wooden hook. The man holding the other end of the hook would draw the shark near, and bash it to death with a chunk of wood.

Will a post-institutional America more closely resemble the relatively secular practice of poisoning reef fish? Or will it be full of taboo and ritual, like shark fishing?

Benassi had an answer.

"Superstitions," he wrote, "seldom supplant or intrude upon rational, empirical approaches, where such approaches will serve."

The lagoon fishermen had large degrees of control over their activities, and therefore did not turn to superstitious beliefs to court success. The shark fishermen, by contrast, had a much lower chance of success, and little control over the factors, and so they turned to paranormal beliefs to guide them.

Sociologists and anthropologists have connected this theory to American baseball players. Jim Ohms, who played for the Daytona Beach Islanders in the mid-sixties, used to put a penny into the athletic supporter of his jockstrap after every win of the season. As the wins piled up, the pennies would clack and jingle every time he sprinted toward a base. When Glenn Davis notched a win, he saved his chewing gum and tucked it under the bill of his cap when he wasn't actively chewing it. Another player kept a lucky cheese sandwich in his back pocket.

But these superstitions, widespread among pitchers and batters, were virtually nonexistent with respect to fielding. The sociologists said that this is because the structure of baseball gives outfielders a high degree of control over catching and throwing a ball, while batters have a lower chance of success as they grapple with the endless mystery of the trajectory, speed, and spin of the next pitch coming their way.

If Benassi is correct, the bad news for institutionalists is that when people feel like they're shark-fishing with chunks of firewood, they will turn to mystical explanations for chaos of the world around them.

However, Benassi found that radically different levels of deference to the spirits may be exhibited by the very same person, depending on the stability of their environment.

This could be good news for institutionalists.

"The occultisms are compartmentalized," Benassi wrote. "We are willing to use astrology in determining our partners for social dating, because nothing else seems to work well, but would not contemplate trusting to the stars if we were making legal and financial arrangements with these same partners."

But Benassi conducted his research before the explosion of the internet. Today, in America, the human tendency to consume the most radical clickbait has buried evidence-based streams of information and accelerated a national atmosphere of extreme political partisanship, perceived instability, a constant fear of disaster, and widespread feelings of helplessness.

With so many people spending so much time looking at the world through this field of distortion, I wonder whether that rationalist center of every spiritualist person—the part of them that goes to college, and fills out tax forms, and assumes that life will continue beyond the next Mayan-inspired doomsday—will continue to hold. I worry that they have become so beset by images of institutional sharks that their actions create something even stranger, and darker. Like Seth Mazzaglia.

Meanwhile, the impacts of institutional distrust on UNH were making it feel increasingly shark-infested.

UNH associate sociology professor Cliff Brown works in McConnell Hall, just a floor down from Benassi. He's been working at UNH since 1996; when I spoke to him in 2023, he was fifty-eight, and serving a four-year term as president of the tenure-track faculty union at UNH.

"I love my colleagues," he said. "I love working through the challenges we face together. But even more than that, I really love my students and I love the moments in the classroom where students really feel like they're, you know, things are clicking and they're getting it and they're participating and they're seeing new things come together. Those are relationships that are deep and meaningful and cultivating those is a real privilege."

Sociologists are important to the broader public, because they produce the knowledge that reduces a city's homeless population, explains the impact of environmental problems on different demographics, reduces child abuse, and makes workforce members feel personal satisfaction in their careers. They analyze data to identify patterns and suss out the root causes of social ills and successes.

"If you've got a problem, say in Manchester with homelessness, you know, maybe it's not just about the people who are homeless, but about the systems that are failing certain categories of people in Manchester or in the state of New Hampshire or nationally," Brown said.

Before the state legislature cut the UNH budget, Brown described the sociology department as if it were an ecosystem awash in mutualism—instructors nurturing incoming students, who became graduate students, who supported those same professors as they nurtured yet more incoming students.

Before the state-imposed budget cuts, Brown taught an introductory sociology course to 130 or 140 students every semester. He accomplished this with the help of TAs, drawn from a pipeline of thirty-five to forty graduate students. These graduate students brought fresh ideas into the department, allowing the department to expand its catalog offerings. A twenty-hour workweek from a TA extends the reach of a faculty member, who can use that TA to help struggling students, and to grade assignments. This allows the faculty to more easily take on roles as thesis advisers, researchers, and mentors.

But the budget cuts had a real impact. Sometime around 2013, the number of graduate students started to dwindle.

"The last three, four years, it's really gone down," said Brown. "Now we have, I think, a total of nine graduate students."

And six of those nine were nearing the end stage of their PhD studies, and were no longer supporting professors in TA roles.

In 2023, Brown was told he couldn't have a full TA. Eventually, he was allocated the money to support a half-TA position, at ten hours a week. That left him, like many professors, to make up the work that the TA would have otherwise performed. A faculty member can either cut back on research, or cut back on student assignments.

"It's a zero-sum game, you know?" said Brown. "We can only provide what I think would have to be seen as a less robust experience for the student."

Meanwhile, Brown's introductory sociology course dwindled to seventy or eighty students, a bit more than half what it once was. In May 2019, BLS stats showed that there were only sixty post-secondary TAs employed in the entire state, comprising a lowest-in-the-nation .09 TA positions per 1,000 jobs in New Hampshire (about a tenth of the ratio of neighboring Massachusetts).

This is just one example of how UNH programs are being hollowed out. Faculty who retire are not replaced. Without any new hires, the department ages and stagnates. It can offer fewer classes, which means larger class sizes and fewer TAs, which means even fewer classes, less demand for services, and, ultimately, fewer faculty. The program's not dead. It's just dying, and the dying process has been going on for a decade. It may go on for a decade more.

"It started as do more with less," said Brown. "Then it became do the same with less. And now, it's do less with less," meaning that ambition for improvement has been slowly throttled from the workplace culture.

The transfer of trust from New Hampshire's evidence-based institutions to spiritual-framed individual truths is leaving those institutions gasping for air. Absent meaningful public funding, UNH is subject to the whims of a market that puts it in conflict with its mission to give New Hampshire students a good education. There is added incentive to recruit out-of-state students, who pay a higher tuition rate. That subtly shifts the student profile, from local students taking advantage of an affordable education opportunity, to students from wealthier families.

"I think we're moving, we're stepping away from that mandate that we serve the students of, of the state of New Hampshire, because our admissions decisions and our recruitment decisions are being driven by these fiscal pressures," said Brown.

During his 2014 annual address, University of New Hampshire president Mark Huddleston said colleges had to resist the urge to chase dollars. "The first thing we shouldn't do is yield to pressures to commodify higher education, turn students into customers, and drive relentlessly to lower unit costs of production."

And yet, that seems to be happening. The programs that are thriving are those that exist to prop up institutions facing existential threats—like UNH's department of security studies, which trains people for careers in homeland security, corporate security, terrorism response, and law enforcement.

As paranormal beliefs grow, some form of this is happening at academic institutions all across America. The question is not whether the tide of institutionalism is receding. It is. The question is: As it does, what new landscape is being revealed?

Twenty-Two

THE FORCE MAJEURE

> REALITY, n. *The dream of a mad philosopher. That which would remain in the cupel if one should assay a phantom. The nucleus of a vacuum.*
>
> —Ambrose Bierce, *The Devil's Dictionary*

"I think the world is going to look drastically different," said Antje. "I think that formal governance is going to look very different. . . . Individuality, individuality within community is going to become the force majeure."

There is little room in this vision for our current world religions.

"If you ask me about religion I believe belief in a higher power is wonderful," said Antje. "Do I believe it in the parameters of the church? No, I do not. I think it was designed to keep control over people. There's no right way. My way is not the right way. It may be for me. It may not be for you. So there's a lot more wiggle room. There's a lot more flexibility."

Modern spiritualists have a vision of a new post-institutional America, one that involves a mass decentralization of power, defunded colleges, empty churches, and a reduced role for hospitals, law enforcement, the scientific establishment, and government.

Antje wouldn't put it that way. She's not targeting institutions, and she accepts the scientific consensus on the value of vaccines, and the dangers of climate change, among many other things. But the major shift in priorities paranormal America envisions would inevitably leave our current institutions behind. Antje feels that science offers a viewpoint into only one facet of a richer universe. In one striking example of how far this could go, she said more people will understand that the physical reality in which we exist is more of a construct than an absolute fact.

"We're loosening our grip on what we believe to be true. Like, I look at this table and it's true that it's there." She knocked on the table between us. "Sure."

But in the future? When fewer people accept the nonintuitive scientific idea of molecular structure—who knows? Tables won't disintegrate, of course.

"That's not what I'm saying," said Antje. "But it's almost like the veil is being pulled aside a little bit and we're seeing the madman, like in *The Wizard of Oz*. We thought it was one thing. But now we really don't have to see it that way."

Antje said this poses a question.

"And if we don't perceive it that way, is it that way?"

Yes! says Cliff Brown, the UNH sociologist. He agreed with Antje, that such a shift would unmoor centuries of hard-won scientific knowledge, and defrock the concept of expertise.

"That just creates this whole world where anything goes, anything's possible," he said. "My knowledge is just as valuable as your knowledge because we don't have a, a shared understanding of facts," said Brown.

While Antje sees this as exhilarating, Brown sees it as harmful. That shared understanding of facts allows, for example, a sociologist to construct a legitimate survey that can achieve statistical significance. If inexpert surveys become the norm, then it becomes impossible to identify the public policy that would, say, best reduce child abuse.

If you value evidence-based institutions, you might be thinking, Fuck. Or perhaps, We're fucked. Or maybe, Why don't these fucking paranormal fuckheads fuck off instead of fucking dragging us all off to Fuckity Fuckville for who knows how fucking long and for what fucking reason?

But it's too early to despair. We don't yet know whether paranormal beliefs can be organized in ways that create new power structures that challenge our existing ones. Conventional wisdom is that to become a transformative political force, spiritualists will need to field political candidates, build vibrant learning and healing centers, and train professional classes of mediums, ghost hunters, exorcists, UFOlogists, and cryptozoologists in ways that will allow them to effectively serve large swathes of the public. A hint of what this might look like arose in 2019, when a national network of activist witches raised money for the Massachusetts Bail Fund, recognizing a kindred spirit between historical witches and those who have been falsely accused by the modern justice system.

Earlier, I mentioned that religious studies professor Diana Pasulka considered UFO beliefs to be an emerging religion, one bolstered by participation from influential people like Silicon Valley bigwigs. Other experts think Pasulka doesn't go far enough.

Ross Douthat, a conservative *New York Times* columnist who has written extensively about the decline of American religion, says that Silicon Valley leaders are heavily influential in what he calls an emerging religion of "all-American pagans," a belief system that stitches together "New Age spirituality, astrology, UFO fascinations, meditation and mind-altering drugs, magic and witchcraft, intellectual pantheism and old-school polytheism and even Satanism."

Organizing spiritualist beliefs into a potent political force could fall to Big Tech, which by 2012 was emerging as one of America's most trusted institutions. Surveys showed that companies like Google and Amazon had more public goodwill than any branch or level of government, the court system, colleges, or organized religion.

Despite Silicon Valley's reputation for hard-nosed rationalism and logic, several facets of the culture make it particularly open to the supernatural: the universal pursuit of visionary innovations leads people to reflexively reject tradition—church attendance in the area is half the national average, while thought leaders embrace personal development through personal gurus, company-sponsored meditation sessions, future-predicting seers installed at company-wide conferences, crystals, tarot readings, pendulums, and astrology. Also, tripping balls.

Douthat said Silicon Valley is infusing "weird metaphysics" into the artificial-intelligence engines that seem poised to define our future, and predicted that, by 2050, all-American pagans will "be seen as a major American religion, not just a minor tendency or impulse."

Others have charged tech bros with essentially creating a metaphysical vision of the future wrapped in science jargon—a techno-transcendentalism in which we must bow down to our superhuman digital overlords, for the betterment of all humanity.

Clay Routledge, director of the Human Flourishing Lab, frames it as a universal human need to find meaning in life. People abandon religion, but not what he calls "the religious mind." In 2017, while a psychology professor at North Dakota State University, Routledge wrote that the inconstant nature of paranormal beliefs makes them a poor substitute for religion. "They are not part of a well-established social and institutional support system and they lack a deeper and historically rich philosophy of meaning," he said. "Seeking meaning does not always equal finding meaning."

Still, if tracks are being laid for a post-institutional world, there is a wide spectrum of possibilities, some of which are far more dire than others. It would behoove institutions to anticipate what that post-institutional world might look like. And to consider how they can best position themselves to survive in that landscape.

One clue lies in deinstitutionalized states like 1980s-era Liberia. In 1982, an eleven-year-old boy was initiated as a high priest by village elders, and then went on to perform black-magic rituals for political factions vying for control of the country. When the country broke into civil war in 1989, he became known as "General Butt Naked," and acquired power by leading a group of nudist militants who fought wearing only shoes and bullet-deflecting magic charms. Under his command, they performed shockingly brutal atrocities that were in accordance with his spiritual beliefs. He instituted a policy that every time they took a village, a child would be brought to him, whom he would sacrifice in order to eat his heart.

In all he killed an estimated twenty thousand people, before converting to Christianity and becoming an evangelist. In 2024, he was in his early fifties, and still active in Liberian politics.

An institution-light power vacuum creates a haphazard, organic process that almost randomly selects leaders from a dark cave, one that autocratic pigs and monkeys might fly out of. My Butt Naked example is just one of many instances in which chaos has produced a warlord.

This is the sort of world where a theater geek and Best Buy worker could assemble a cult of mindless warriors and commit acts of violence in the name of entities from beyond. But a spiritualist-dominated future doesn't have to leave us in thrall to AI-worshiping siliconeheads, or to cannibalistic warlords, or to Geek Squad rapist murderers. In the West African country of Côte d'Ivoire, we find a more hopeful example of post-institutionalism in the form of *dozos*, according to an analysis by Joseph Hellweg of Florida State University. In some African regions, *dozos* are traditional hunters who perform occult rituals like animal sacrifice as part of a spiritualism that also gives them a role in healing and in resolving disputes among members of the public.

Dozos have long protected people from threats like shape-shifting sorcerers in the forest, a soul sickness called *nyama*, and more mundane local fauna. If an agricultural field becomes a place to force an elephant, or even an animal like a bat out of, Hellweg says, *dozos* have long filled the role. In the 1990s, Côte d'Ivoire went through a crisis in which governmental security forces declined. In the power vacuum, *dozos* became a security-force and conflict-resolution go-to, Hellweg said.

But is our future more likely to be dictated by a murderous Butt Naked, or by benevolent *dozos*? Luckily, we have some insight into what cultures and values are likely to be at the forefront of an institution-light American future. That's because in 2023, Pew began tracking the 22 percent of Americans who have taken up the emergent identifier of "spiritual, but not religious."

While spiritualists hail from all demographic sectors, there is a sort of "typical" profile of people who are spiritual, but not religious: a white Democratic woman under fifty, who is more likely to have a crystal than a cross, and who has had a personal experience with an otherworldly presence. Generally speaking, they believe that people, animals, and even geographic features, like a mountaintop, have spirits that can talk to and physically affect the living. These spirits can be either helpful or harmful to living humans. They don't draw their morality from the carrot-and-stick visage of Heaven and Hell; instead, they are more likely to have an organic morality influenced by their tendency to spend time in nature, and by their belief that being connected with nature is an essential ingredient of their spirituality.

So perhaps we will instead be in thrall to a legion of stereotypical Earth mothers, whose spiritual influences will fill the roles currently occupied by priests, scientists, professors, and therapists.

If Beau and the others represent one possible future for the paranormal movement, then another likely candidate is embodied by Andy, who was so determined to unify the scientific method and the spiritual realm.

If KRI were to establish a scientific basis for spiritual beliefs, one way to look at it is that it would legitimize spiritual beliefs. Another viewpoint is that such beliefs are already legitimized, and that it could preserve science, and an agreed-upon set of facts, for future generations.

One irony simmering beneath the surface of the paranormal movement is that the triumph of the science wing would create the sort of strict definitions that are anathematic to the spiritualist wing; if there are agreed-upon facts governing the afterlife, individual experience would no longer be the prism through which the spirit world would be viewed.

By late fall of 2012, there was snow on the ground, but no one had shoveled or sanded KRI's area of the plaza parking lot, nor the sidewalk that led up to the front door.

Antje noticed. Her legal practice was rife with disputes that began when a careless heel met the too-slick blacktop of a careless heel, setting off a cascade of events that ended with mass amounts of money changing hands. She wanted something done about the sidewalk.

She reviewed the leasing agreement with the plaza's owner, and found that snow removal was the responsibility of the tenant—in this case, KRI.

But then she saw another clause in the lease, a requirement that the tenant carry renters' insurance to guard against various sorts of accidents. She went to Andy. Was KRI's insurance up to date?

Insurance, Andy told me later, is a very suspect concept.

"I think it is morally wrong to rent your car, your apartment, from an insurance company. Every dollar you give them, they keep half," he said.

Andy translated this ethic into practical KRI policy. He was willing to court risk, even risk court. "One of the not-so-legal things we did was the lease says we have to have insurance and I never did." To Andy's deep satisfaction, no one ever checked. Until Antje. The two sparred over the insurance policy in jocular tones.

"There's Antje!" Andy would say, when they were all together. "Toeing the line, and doing her thing!" If someone slipped, they couldn't sue him anyway, he said, because he didn't have any money.

"Jesus!" Antje would say, expressing mock-horror at the idea.

The jocularity was only on the surface. Andy's joking dismissal of Antje barely hid his true dismissal of Antje. Antje's mock-horror just barely concealed her actual horror.

"You know how some people are really intelligent but they have no common sense? That's Andy," Antje later said. "He was soooo . . ." she waved her hands to indicate chaos and frustration.

"I'm an extremely organized person, so that made me crazy," she said. "I would see some sort of loophole and try to cover it liability wise. Or I would see some risk and say we need some insurance. What are we going to do? What if somebody falls down? He would make fun of me for doing it."

But Antje was a litigation paralegal. She was extremely uncomfortable with putting her name on an organization that was flouting the law.

"She freaked out," said Andy. Adding a $1,000 expense to the bottom line of the Center, which was just barely keeping afloat, was just out of the question. Over the past year and a half, KRI had lost money for every month it was in operation. "A thousand a year doesn't sound like a lot but it's more than I have," Andy said.

During these disagreements, Val often felt like she didn't have enough standing to weigh in. Beau mediated, trying to bridge the gap between them. Mike was unfailingly supportive of everyone, but tended not to take sides. Mike, incidentally, seemed to be burdened by his personal life. He had gone through a divorce from his wife, Donna; had put on weight; and sometimes looked pale. He moved in with a cousin in October, and then moved into his own place, a Spring Street apartment, the following month. His drinking seemed to be on the rise.

Andy's resistance was especially irritating to Antje, because ultimately, everything was in his name, under his personal guarantee. He was the one she was ultimately protecting. Antje offered to do all the legwork—using her expertise, she would shop around, fill out the forms. When Andy said he didn't have the cash, she took the extraordinary step of offering to pay for it out of her own pocket. Then she did.

The insurance issue turned out to simply be a proxy for a much larger personality clash between the two. Andy was, if not happy, then go-lucky. Antje was, if not a Johnny, then a come-lately, which hampered her ability to have her legal advice implemented. They had a, well, to-do.

Antje was used to negotiation, but Andy never wanted to engage on that level at all. Antje wanted to build the psychomanteum. Andy wanted to finish the retail space, and do other physical improvements to the Center. They had a merry-go-round over priorities.

"Her complaint isn't that I overfocused. It's that when I was focused, the things she asked for were put in second place," Andy said. "The psychomanteum is really a good example of that. My frigging shower is going to work before your psychomanteum."

Because Andy was the undisputed decision-maker, Antje relied on soft power—influencing him through argument, through repeated requests, and through Beau, who was increasingly caught between two good friends.

"I felt abandoned by Andy," said Antje. "And I was pissed. And he got very defensive, and said, 'My way or the highway, and if you don't like it there's the door.' And I felt very dismissed. And very unheard. And it was a shitty feeling."

"It's not like I hated her," said Andy. "You know how you get that little dog going yip yip yip? You want to swat it? That's her to me. It's a cockapoo. I know if I give it a bone it will shut up for half an hour. I had no respect for it. She was so far up my butt that I surrendered a lot of the decision-making processes to Beau, just so she would stop annoying me."

On Sunday, December 2, police officer Ernest Orlando was watching the time from the front desk of the Farmington Police Department. It was 10:30 at night, and the public was, for the most part, going to sleep.

Then the doors to the station opened, and a burly man with thinning hair and a thick black beard entered. He wore a pea-green T-shirt and his arms were heavily tattooed—two yin-yang symbols, a wolf, two moons, and letters. It was Mike Stevens.

He said he wanted to turn himself in.

Orlando asked Mike if there was a warrant out for his arrest. This was a fairly routine occurrence.

"No," said Mike. "I killed my wife."

"You killed your wife?" Orlando repeated, startled.

Mike said that he had strangled her.

The officer asked Mike to step into the booking room. When he did, Orlando noticed the smell of alcohol in the air.

Twenty-Three

THESE WEREN'T HUMAN DOCTORS

He reverences truth, he loves kindness, he respects justice. The two first qualities he understands well enough. But the last, justice, at least as between the Infinite and the finite, has been so utterly dehumanized, disintegrated, decomposed, and diabolized in passing through the minds of the half-civilized banditti who have peopled and unpeopled the world for some scores of generations, that it has become a mere algebraic x, and has no fixed value whatever as a human conception.
—Oliver Wendell Holmes, *The Poet at the Breakfast Table*

It was a Tuesday in late October 2013. Eleven months had passed since Mike Stevens went to the police station with a confession to a horrendous crime (and a whiff of alcohol) on his lips. Mike was saddled with the trauma of his abduction, but tonight, the KRI organization was focused on a different trauma, a different abduction.

A former logger named Travis Walton stood near the front of the room. Nearing sixty, Walton was tall, with thinning red hair, a mustache, and an erect posture that suggested discipline. His black suit, crisp blue shirt, and tie made him look like a seventies-era salesman pitching an amazing kitchen gadget to a television audience.

In the 1970s, as Betty and Barney Hill's fame waned, it was Walton who had picked up the baton of most popular abductee. His case was so famous that he had been portrayed by actor D. B. Sweeney in a 1993 Hollywood movie called *Fire in the Sky*.

He was, by far, the most high-profile presenter in KRI's history.

"It was a big deal," said Val. "It was a very big deal."

Val herself had undergone a transformation over the past year. It began when she was spurned by Antje. She'd heard, through Andy, that her negativity was off-putting to both Antje and Beau. It prompted Val to look within, working to understand and defang her own internal darkness.

"My complaining wasn't selfishness," she later said. "I just wasn't aware. I didn't realize how toxic it was."

At first, it was difficult to stifle her first instinctual response, but over time, she found different things to say, things that were positive and, she was happy to find, also true and real parts of herself. It was like excavating a long-buried joy. She unearthed that little girl who had come to kindergarten expecting the best of the world.

Beau and Antje did not rush to embrace the new Val, but that was okay. She admired them, but no longer pined for their attention.

"They were not the type of friends who were willing to help me by telling me that I was in a negative place," she said.

And so, when Travis Walton agreed to come to KRI to screen a documentary about his abduction (Antje had paid several hundred dollars for the big projection screen), this new Val, brimming with positivity, was upbeat. She relished this rare opportunity to meet a man who had spent five days aboard a spacecraft from another planet.

They sat in the dark and watched his story. One evening in 1975 in Turkey Springs, Arizona, as Walton's work crew drove off a logging site, they saw a lighted object among the treetops. Twenty-two-year-old Walton, even then a UFO enthusiast, exited the truck and ran toward the object, only to be swallowed in a massive beam of light that shot down at him from above. The driver of the logging truck turned the vehicle around and floored it, taking the rest of the crew with him. When they returned, fifteen minutes later, there was no sign of the light, the object, or Walton. Over the next couple of days, search parties failed to find him in the woods. Five days later, after the search was called off, Walton called his family from a roadside pay phone that was twelve miles from the abduction site.

During the ensuing media blitz, Walton reported that the beam of light had knocked him unconscious. He woke up lying on a table beneath a bright light, tended by doctors. His eyes were difficult to focus.

"It was only a matter of seconds before I realized that these weren't human doctors," he later said. Walton jumped up in distress and grabbed a glass instrument from a table, trying to break it to fashion it into a weapon. When the aliens fled, Walton wandered the spacecraft, noting that he could see stars through the walls. A human wearing a spacesuit helmet appeared and escorted him into another room with a glowing ceiling and two other humans in it. Walton asked the three of them questions, but they didn't speak. They put a clear plastic face mask over his mouth, after which he fell asleep. He woke up lying on the pavement, and looked up just in time to see a bright light shoot up into the sky, and disappear.

After the experience, Walton seemed dazed, and only regained his sense of composure from the passage of time, and a hypnosis session that allowed him to recall the details of his experience.

Andy once told me that there were far more reported alien abductions than could actually possibly be happening—estimates range from several hundred thousand to 3.7 million abductions in America alone. Andy felt that, though many abduction stories were the results of alcoholic blackouts and other mundane factors, the Travis Walton experience was a leading example of an authentic abduction.

Andy's word was carrying more weight these days, because in May he'd earned his BA from Granite State College. Hoping to be able to perform research that would carry weight in the academic world, he was now pursuing graduate studies at UNH.

Andy was following his mentor, Bill Stine, into the field of psychophysics. But while Stine used psychophysics to help explain paranormal beliefs as baseless superstitions, Andy planned to use psychophysics to legitimize those same beliefs. Once Andy understood that Stine wasn't open to paranormal phenomena, he never mentioned them again. There was no mention of ghosts in his academic project, "The Effects of Spatial

Frequency on Local and Global Stereo Acuity," but Andy hoped to pin down aspects of audio processes in a way that could shed light on whether ghostly manifestations—the shimmering form, the whispered voice, the skin-prickling chill—were physical effects, or existed purely in the sensations of the experiencer.

"This, specifically, is why I'm doing this field of study," Andy said. "It's a methodology. There are a lot of things that I can do using this shared methodology."

Walton's skeptical critics pointed to sketchy circumstantial evidence. Shortly before the event, Walton had told his mother not to worry if he was abducted by aliens, because he was sure he would be returned; more tellingly, his logging crew was just days from defaulting on its federal logging contract because they had not done enough work on the site. Walton argued after the abduction that an unforeseeable act of God had occurred, and he should be released from the contract.

But at KRI, Walton was greeted with credulous arms. Andy found Walton's abduction story to be the most credible one he'd ever heard. As the documentary showed re-created images aboard the spacecraft, Val watched Walton grip the arm supports of his chair, the knuckles of his hand going visibly white in the shadows.

"To find him so down to Earth, doing what he was doing with the fears of his encounter and still talking to other people, that was very touching to me," said Val.

Though KRI didn't charge a gate fee, there were new faces in the crowd, and the donation box was full. Andy (who qualifies that he is "not a records guy") said he thought this was the first month that KRI took in more money than it spent. After years of scrapping, it was beginning to look like KRI might actually make it.

KRI posted a picture from the event on Facebook. In it, Walton posed with a fellow abductee, the two of them smiling and holding one another in a show of abductee solidarity.

It was Mike Stevens.

Mike was the one whose connections within the UFO believer community convinced Walton to come to the Center. When the lights dimmed and images of a spacecraft danced across the projector screen, Mike was there, watching silently. But at least some part of his mind had to have been occupied by the criminal charges that were hanging over his head.

They stemmed from the night that he had walked into the Farmington Police Department and announced that he had killed his wife. Or maybe she had survived. He wasn't sure.

After Officer Orlando walked Mike back into the booking room, Mike repeated, in front of Sergeant Ferguson and Lieutenant Krauss, that he had strangled his ex-wife. Ferguson heard Mike say, "I think I choked my wife too hard." But he also said that he thought she was conscious when he left. It was hard to know for certain whether a murder had taken place, but certainly it seemed that something had happened.

Orlando told him that, until they figured out what was going on, he wasn't under arrest. But he would have his rights read to him before he provided a written and audio statement for the record. Mike agreed. While Orlando quizzed Mike in front of an audio recorder, Ferguson and Krauss hurried over to the scene, a rear apartment on Grove Street.

There, they found Mike's ex-wife, Donna. Her throat was red and raw, and the blood vessels in her eyes were abnormally visible, but she was still very much alive.

As they were interviewed—Donna on her couch, Mike at the police station—each gave similar accounts of what had happened. Though they were separated, they spent the day together at Mike's place on Spring Street to celebrate their daughter's twelfth birthday. Things between them were tense, but they kept it together throughout the party.

Donna left the party early, at 6:30 p.m., with her cousin. Mike called her and asked her to come back to the party, but she refused, and stopped answering repeated calls from him. They began arguing via text. When she turned her phone off, he began texting their daughter on her iPod Touch, telling her to wake her mother up. Donna turned her phone back on and told him that they would discuss it tomorrow, when he was sober. Mike asked her why she didn't love him anymore. She hung up on him.

Mike drove to the apartment and let himself inside. Donna, in the hallway area off the kitchen, shouted at him to leave and called him drunk. He said he was going to kill her. He put his hands around her neck and squeezed.

Donna's son told authorities he came out of his bedroom and saw what was happening, but was too shaken to do anything other than return to his bedroom. Their daughter reported hearing the shouting but stayed in her room.

Donna was forced up onto the tips of her toes, and started to see spots. When she began to struggle for breath, he let go, slapping her on the left cheek and knocking her to the floor. He turned on his heel and walked out.

Donna grabbed her inhaler, then went into the bathroom, where she cleaned herself up and washed her face. She seemed unsteady, and told the officers that the left side of her neck hurt.

A Farmington ambulance showed up, and the EMTs said they would escort Donna to Frisbie Memorial Hospital (a besieged institution that would soon close its labor and delivery unit) to evaluate her injuries. Mike's mother came to look after the grandkids while Donna was away.

To sum up what he had told Officer Orlando, Mike wrote a confession on a mimeographed police statement form. "I got mad over situation," he wrote, "choked Donna, told her I should let her die but wouldn't, open hand smacked her hard and left."

While sitting with Mike, Orlando's phone rang. It was Ferguson, summarizing what he had learned from Donna. After a brief exchange Orlando hung up the phone.

"Mike," he said. "You're under arrest."

Mike said he understood. When he learned that Donna had not called the police, he said he was surprised. Orlando asked if he had been drinking.

Beer, said Mike.

After nearly two weeks in county jail, Mike posted $1,000 and was released on bail. Donna declined to press charges, but prosecutors had her statement, and physical evidence of assault. Assistant County Attorney Amy Feliciano began to build a case against him.

In January 2013, Mike pleaded innocent, and in February, he was indicted on a felony assault charge. The prosecutor began to line up various witnesses, and convinced Donna and the two kids to give additional statements about what they had seen. The prosecutor also secured permission to access Donna's medical records from the night of the assault.

When fellow abductee Travis Walton came to KRI to talk about his abduction, Mike's deadline to commit to a guilty plea, or face a trial, was about to run out.

Within the KRI community, the controversy surrounding Mike's court case was playing out along familiar lines—Did they trust the individual, or did they trust the institution that was prosecuting him?

Beau, whose spirit-informed wisdom was gaining her a bigger following than ever, said her relationship with Mike actually deepened during this period.

"I am the first person to say if you put your hands on a woman, you should be held accountable," she said. "What Mike was accused of doing, he didn't do. He's the yahoo who couldn't step on an ant."

In 1993, Gallup polling found that more than two-thirds of Americans felt that the US justice system treated those accused of a crime fairly. Since then, America's justice system has been tattooed with the ink of its failings—for me, the critical moment came when I witnessed the emptying of innocent men from death row in Illinois. But that's just me. For other Americans, it was George Floyd's murder, Rubin "Hurricane" Carter's incarceration, the pardoning of January 6 insurrectionists, or Bill Cosby's insulation from prosecution. More broadly, American justice has also been stained by a failed war on drugs, a general sense of overcriminalization that has resulted in an estimated one in three adults with a criminal record, an upswing in the use of military gear and tactics, police brutality, and of course the systemic biases that disproportionately target the poor, and people of color, with the net effect of expanding America's wealth gap.

And so, it is perhaps unsurprising that by 2023, fewer than half of Americans believe their justice system treats the accused fairly. In other words, more people believe in aliens (57 percent) than believe in the fairness of criminal justice outcomes (49 percent).

Two months after hosting Travis Walton at KRI, on January 2, 2014, Mike pleaded guilty to misdemeanor charges of simple assault and criminal threatening.

Andy, Antje, and Beau were in the courtroom to support Mike.

Analytical Andy viewed the court proceedings as illegitimate.

"It was a joke. It was a farce. It was insane," Andy said of the charges.

Empathic Val had no feelings of foreboding. "I never felt afraid of Mike," she said. "The story never troubled me. I knew his core, his heart, well enough to know he's not a scary, violent guy by nature."

And risk-averse Antje said she would happily open her home to him, no questions asked. "I trust him implicitly, and I've never known him to be a violent or dangerous person," she said. "Quite the opposite."

The judge agreed to a deal that gave him twelve months in a local house of corrections, and a $2,000 fine, with a chance for work release after ninety days.

But Mike was led behind bars to serve out his term. He'd been abducted, not by the aliens in which most people believe, but as an expression of the law's fairness, in which most people do not.

Twenty-Four

THE SCRYING MIRROR HUNG

It's money in one's pocket these days to be dead, for ghosts have no rent problems, and dead men pay no bills. What officer would willingly pursue a ghostly tenant to his last lodging in order to serve summons on him? And suppose a ghost brought into court demanded trial by a jury of his peers? No—manifestly death has compensations not connected with the consolations of religion.

The marvel is that apparitions were so long in realizing their possibilities, in improving their advantages.

—Dorothy Scarborough, *Humorous Ghost Stories*

When I ask Antje about her relationship to forms, she waxes rhapsodic. The word *fanatical* springs from her lips.

Filling out forms, creating forms, handing forms to others to be filled out and stapled to yet more forms, forms in duplicate, in triplicate, forms photocopied into a thousand identical iterations of precision and order. To Antje, it was all magical.

After she joined KRI and before Mike was convicted, she emitted a geyser of paperwork that covered every square inch of the offices, and of every person who ventured inside—class participants were asked to sign on to a newly drafted KRI Code of Conduct, in which they agreed to leave their cellphones at the door, keep intuitive information to themselves, exhibit "a positive and caring attitude," and express "opinions without demands, complaints, and criticism."

She also created a release form that allowed KRI to use photos of their event attendees; an application to host an event at KRI; a waiver of liability for the Psychic Sampler; a waiver for ghost hunters to sign (and another for property owners); a boilerplate legal agreement for guest event presenters; corporate registration with the State of New Hampshire for KRI; and a separate corporate entity for "Continuum Gifts," the retail wing (a separate consignment agreement established that people selling goods in the retail space would receive 40 percent of the sale price, or 50 percent of the sale price in store credit).

Every form was part of Antje's effort to provide clarity in an area that is typically legally gray. Shouldn't ghosts be treated like every other potentially dangerous force in the world?

If contacting spirits is a giggly hobby, then no. If it is a serious business, then yes. After all, she reasoned, if spirits are indeed real, then they have the power to hurt someone more seriously than the wobbliest staircase, the iciest sidewalk, or the hottest McDonald's coffee.

Antje was also creating a network of forms and regulations governing usage of the psychomanteum. They were on the cusp of being used for the first time because, after roughly a year of preparation and construction, the psychomanteum under the stairs was finally complete.

In addition to making KRI a destination, and in addition to providing an avenue of interesting research, the psychomanteum also had the potential to be a big moneymaker for KRI. They planned to charge $100 for the first half hour, plus $25 for each fifteen-minute block thereafter (about the same rate that Beau charged for a reading). Even a modest month of booking had the potential to pay for the entire combined $1,600 rent for the Center and the retail space.

Speaking of which, Andy was still working on the retail space.

The Center had opened in the spring of 2011. It was now the late fall of 2013.

"He did everything for the store very slowly," said Val. "He had that space rented for a really, really, really long time and it was in the works the entire time. He was handcrafting every piece of it and wasting time was how it felt. Why don't we buy some cheap-ass bookcases and get it up and running? It was definitely very frustrating."

But Val was not one to rock the boat. She was still working on being less negative—if there was a positive change in how people were relating to her, it had not yet undone a lifetime of caution about stating her opinion.

"I hadn't found my voice yet," she said. "I would be frustrated but didn't verbalize it most of the time."

When Andy disappeared into the retail space to work, it was as if he were stepping through a portal into another dimension, a place where the headaches and anxieties of keeping KRI afloat faded into the background. While he tapped nails into place and sanded wood, hours slipped by.

And yet, his blissful reveries could not forever forestall the intrusion of a thought, one that began as a small soft supposition that hardened into an ugly nugget of unwelcome certainty. The retail wing of the center, he realized, would never come to fruition. The cost to stock it with inventory was hopelessly out of reach. And even if they had stuff to sell, there was no way to staff it. It was as if gloomy crows had alighted on his shoulders, whispering these words of doubt. And yet, he didn't tell the others. He didn't stop working. He just kept on crafting. The counter was nearly finished now. He'd bought lumber to build a frame around some scavenged glass; the display case had beautiful lighting that would never shine on healing crystals, gleaming shelves that would never bear the weight of commerce, a bar-back lifting section that would allow nonexistent staff to enter and exit the behind-the-counter space swiftly, to help the nonexistent customers in between spells of tidying the nonexistent inventory.

There was a reason that it was taking Andy so long.

If the building owner had been paying attention, he might have noticed that the ceiling tiles Andy was installing in the retail space were a bit lower than they should be. That the landing on the stairs leading to the balcony level had a pile of lumber and some drywall without any obvious use. That he'd hooked up grounded outlets that allowed for more electrical support than the retail center would realistically need. That he was doing something to the plumbing?

The explanation was simple. Andy was sick of cramming himself into the tiny KRI bathroom to do his dishes and shower in a bucket. He was building a secret apartment into the retail space's second floor; the dropped ceiling tiles would allow him to angle the bathroom pipes enough to satisfy state building codes. For the first time since the turkey coop, he would be able to take a real shower.

The fact that the owner hadn't noticed and intervened was not surprising to Andy. The property manager had a tenant-management philosophy that Andy summarized as "Don't tell me, and don't blow anything up."

So Andy kept toiling. Though Walton's appearance had led to a profitability blip, they remained in dire fiscal straits.

Which is why, when the psychomanteum was completed in December 2013, everyone was excited. Beneath adjustable wires, a scrying mirror hung, a nexus between worlds, awaiting its first use. No one knew how successful it would be. Such an unusual offering could even receive some attention from the mainstream media.

I wasn't there, so I can't say whether Antje literally licked her lips in anticipation of drafting documents for the psychomanteum, but certainly, the prospect quickened her blood. Her psychomanteum application form established that those who were taking medication or seeking treatment for any mental-health disorder, or any physical condition that might be affected by stress, had to submit a letter from their provider granting specific permission to use a psychomanteum.

She combined a liability waiver with written rules and regulations that screened out high-risk clients, including those who, she wrote delicately, "have less than a full grip on reality." Also considered high-risk: those in the throes of extreme grief from bereavement, and those who were under twenty-five years of age. Users agreed to protect themselves spiritually through prayer or the equivalent, to abstain from drugs and alcohol for at least twenty-four hours before entering the psychomanteum, and to avoid intentional contact with negative entities. They also agreed to leave their cellphones, recording devices, and shoes outside, and to assume full risk with a witnessed signature.

The Scrying Mirror Hung

Antje's final psychomanteum document was a set of protocols for session facilitators, designed to make the data and reported outcomes more uniform, and therefore scientifically defensible. The facilitator would create a new file for every session. For each, they would ensure that the payment and paperwork were in order, guide the sitter through a period of meditative focus on whom they would like to contact, and then ensure that those spirits were crossed into the afterlife. The facilitator would then escort users into the small space, to settle them and give them instructions. Later, the facilitator would give a two-minute end-of-session warning, and then debrief the user. This would allow the Center to accumulate data and common experiences.

Antje asked facilitators to walk a fine line, by ensuring that the clients understood the dangers of parasitic entities and evil demons, while keeping it fun.

"Keep a sense of humor about what you are doing and allow that humor to live in the psychomanteum," she wrote.

Finally, it was time for the first user. Antje, her heart trilling in excitement, ducked into the psychomanteum she had created, at an out-of-pocket cost of about $1,200. Thanks to the soundproofing, and the rippled blackout curtains that Antje had commissioned from a seamstress, it was like sitting in a black hole in the galaxy's outer rim.

You can't hear a damn thing.

The thought gave her satisfaction. There was no outside light, no smells. Just her own breath, her own heart. She sat in the zero-gravity chair, let her eyes adjust to the paucity of light, and gazed at the mirror, which was angled so that she could not see herself in it. This was a scryer's dream. Antje relaxed her body.

She could relax, in part, because she knew the psychomanteum was also a perfect cocoon of safety. Raymond Moody had recommended that Antje use a candle. But a candle was a fire hazard. So instead, she installed a single battery-operated tealight, positioned behind her head, flickering. The flickering had the potential to induce a seizure, and so Antje had specifically screened people with seizure disorders out of the pool of users.

Antje stared at the center of the mirror. She could faintly make out images of the rumpled black curtain edges behind her. Then she looked beyond the center of the mirror, to its far side.

The images blurred, and became milky. Every time she blinked, the milkiness went away, but came back. She wanted this milkiness to turn into a billowing mist.

"They may discern shapes, faces even, in the mist," she wrote to facilitators. "With practice and experience, these shapes can become so real that they seem solid. Indeed many psychomanteum users claim to have physically held their departed loved ones in their arms for a while, having come out of the mirror to them. Other users claim that they went into or through the mirror to meet other beings."

As she sat, blinking in the darkness, Antje was mindful that her first session might not be successful. To acquire the practice of defocus takes time. Even Moody had spent a year before he saw anything.

"I wasn't expecting it to be like, 'Whoa, my grandma climbed through the mirror and said hi.'"

And boy did she not.

Antje saw nothing. That was just fine. The calendar of possible sessions stretched out before her. She would do it, no matter how long it took.

That's when she heard the news from Andy. The Center was closing.

Well, maybe closing. Andy told the other members of KRI that he had answered a knock at the door.

Outside was a stiff-necked man who informed Andy that he was the new property manager. Ownership of the plaza had changed hands, and so the old, soft property manager whose live-and-let-live bonhomie Andy had so enjoyed was gone.

The new guy was about Andy's age, but he acted older.

"He was definitely more officious," said Andy. "Not particularly bright."

At the man's request, Andy escorted him around the rental unit. They wandered into the retail space, and the man's forward progress ground to a halt at the bottom of the stairs leading to the balcony level. Andy's drywall sheets and two-by-fours—the bones of his dream apartment—were clearly visible on the landing.

For agonizing moments, the man continued his conversation with Andy, who studiously avoided mention of the second-floor residential unit that he was in the process of building into the commercial space.

Finally, the new property manager allowed himself to be led away.

"He had no idea what it was there for," said Andy.

But Andy's relief was short-lived. Two months later, the man knocked on the door again.

"Andy," he said. "You can't live here anymore."

Andy explained that his presence on the property overnight didn't do any harm.

The property manager told Andy that it was illegal for him to live on the property.

Andy tried to explain it to him. It wasn't illegal. It was just against the lease. The lease might say that the space was to be used for business purposes only. But that didn't make living there illegal.

Things got heated, and began to spiral. By the time it was over, they appeared to be at an impasse.

The property manager wouldn't tolerate Andy living at the Center, and Andy wouldn't tolerate living anywhere else. He had the renters' insurance! What more did they want from him?

"That's the moment the realization hit," Andy said. "The ability to finish that store, with an apartment over it, was done."

Though the argument between Andy and the property manager dragged on for months, it was clear they would have to abandon the building soon. Antje was crushed. What about the psychomanteum?

"It sucked," she said. "It sucked."

Before it was time to move, Andy and Val gave the psychomanteum a try, but didn't see anything. When Beau settled herself into the chair, her spirit guide showed up—not in the mirror, but off to the side.

"My guide was like, 'Why?'" It was the table tipping all over again. "Because I can already see them."

While Antje maintains that the empty mirror was simply proof of the need for practice, Val and Andy had a different take. Toward the end of Val's forty-five-minute psychomanteum session, she grew impatient, and tried to tap into her spirit guides for advice. That's when she realized that she couldn't tap in. There was an energy field there that thwarted her.

Oh my God, Val said to herself. *Nothing can get through.*

Antje's forms, she concluded, not only protected KRI from liability—they also protected psychomanteum users from spirits of any sort.

"Her fear of someone going in and using it for the wrong reasons was preventing anything from happening," Val said. "That created an energetic barrier."

Andy agreed.

"In her desire to make everybody participate to be safe, she made them safe from everything," said Andy, "including what they wanted."

And just like that, the psychomanteum experiment was over. It took Antje a year to put it together. Now, it was time to take it apart.

Twenty-Five

A STAKE IN THE CHAOS

> PANDEMONIUM, *n*. *Literally, the Place of All the Demons. Most of them have escaped into politics and finance, and the place is now used as a lecture hall.*
> —Ambrose Bierce, *The Devil's Dictionary*

As they had done when Andy needed a new car, he and Beau went on a search, looking for a new home for the Center. But, as she led him to a variety of places, Andy said that his air of expectation was replaced by disappointment.

"She was taking me to places that were in no way suitable," Andy said. "Nothing about her choosing places made sense. It was the complete opposite of what happened with the car."

Some were wildly out of his budget, and others wouldn't allow Andy to live there, which was, to him, a nonstarter. She found him an old medical center with a beautiful cathedral ceiling, but he said it was too far off the beaten path, miles deep in the woods. "Nobody's going to be able to find it," he said. One place had an apartment, but no air conditioning. Another nonstarter.

Beau, for her part, was starting to hear Antje's complaints about Andy. Some of his personal peccadilloes began to feel problematic. The hours he had sunk into renovating the retail space now seemed particularly counterproductive.

Other issues were cropping up. Antje, freshly hurt by the loss of the psychomanteum, said Andy only cared for himself. Val and Beau said he was bad at delegating. Everyone agreed that a website was needed, but Andy wanted to do it himself, and it was perennially on the back burner. Andy, for his part, was frustrated by what he saw as inadequate buy-in from the other members, who were, in the end, volunteers that he could not compel to do things.

With all this negative energy swirling around, none of them realized that there was something in the Center that had a stake in the chaos.

KRI decided that, before they lost access to the retail space, they should do a final in-house ghost hunt of their combined rental space. They knew by now that the Center was infested with spirits—guides who had made a home, benevolent beings that played a supportive role in their various spiritual activities, as well as a rogue's gallery of unknowns and ne'er-do-wells. Who had banged on the rooftop to frighten Val and a friend one night? Who had toppled that raccoon?

And so, with a couple of trainees in tow, they undertook a census of their spiritual assets and responsibilities. That night, they broke into smaller groups headed by the three intuitives—Val, Antje, and Beau—and walked the corridors and office spaces that seemed so different in the daylight, listening for murmurs from beyond.

As Val picked her way past the tools and materials of Andy's unfinished retail space, her stomach began to churn in a way that was unpleasant, but familiar. She was picking up on an energy signature of a malevolent being. Val had encountered this many times before, in houses or even while driving past a vacant lot.

By the end of the night, the others would confirm that they, too, sensed the malevolence lurking in the retail space. Soon, they had identified it as "an entity that fed on negative energy."

A parasite! It had been hiding in the Center, right under their noses, though it was difficult to say whether it was actively sowing negativity, or just feasting on things as they were.

"Things like that tend to go where humans create negative energy," said Val. "We tend not to need help producing negative energy."

No one knows exactly what a spiritual parasite is, how many there are, or what they can do. Some think that packs of them latch onto a single individual, feeding off their psychic energy. Others associate them with demons, and others with poltergeists. Most people agree that a parasite can be dangerous, because it has an agenda of increasing negative energy, which can drive a human host into depths of despair, fits of anger, and full-on panic or terror. Every bicker was a snack; every complaint an aperitif.

"And looking back," Val said, "it made sense that it was in the retail space where Andy was having so much trouble moving things forward. He may have been affected by it."

This was where Andy, who famously did little to protect himself from spiritual dangers, had become an eternal worker in the model of Sisyphus, laboring on a project that no longer had any purpose, while his friends in KRI complained about his absence.

When Val first sensed the parasite, she knew that they'd be able to deal with it. She didn't say anything right away, though. In keeping with the triple-medium protocol, she would record her impressions individually, and then compare her notes with Antje and Beau at the end. And Beau would take the lead in figuring out how to get rid of it.

And so, when Val led her small group out of the retail space, she didn't disclose her findings to anyone. As Antje, who was leading another group, entered the retail center, she didn't have any immediate impressions. If there was an entity there, it had retreated from her perceptions and crouched in some small space, hunched and watchful. Antje tried to open herself up; moment by moment, she was widening her lens, scanning the space around her for that which her eye could not see. One of the trainees was a big, slow, sort-of-bumbling guy who looked wonderstruck at every hint of paranormality. It was unclear whether he had what it took to be a ghost hunter. He was hapless, seemingly unable to tie his shoe without triggering some drama or calamity. But perhaps he could acquire the skills and dedication the role required. After all, in this Center of bizarre phenomena, stranger things had certainly happened.

When it was time for the trainees to investigate the retail space alone, they split up in the room, and moved silently through the blackness; Andy's half-finished creations loomed.

The senior KRI members outside the retail space were jolted by a piercing cry—not a parasite, nor a banshee. A trainee. When Antje and the others entered the room, they turned on the lights and rushed to the bumbling trainee. He had been trying to walk over a stack of lumber that Andy had purchased. A nail had pierced the sole of his sneaker, gone right through the meat of his foot, and poked out among the shoe's laces. They stared at the blood pouring from both sides of the stigmata.

Andy and Antje began a quick, low-toned argument over who should drive the trainee to the hospital. The group decided (over Antje's objections) that Antje should drive him to the hospital. Antje said that, since Andy had invited him, Andy should do it. But when Andy said he didn't have any money to cover the hospital bill, she reluctantly agreed, and when she got him there, she pressed several hundred dollars into his hands to cover his hospital bill, and offered a silent prayer to the spirits that he wouldn't sue.

Back at KRI, the others called the investigation off.

They agreed that, on moving day, the parasite, too, would have to be dealt with.

After an exhaustive search, at the end of March 2014, Andy and Beau finally located a home for KRI. They'd settled on a business plaza on Portsmouth Avenue in Stratham, only a couple of miles away.

Before their planned move-in on April 1, Beau arranged to paint the new Center. She selected the colors, and one day, a few of them showed up and began painting. But within a couple of hours, before the work was half-finished, everyone had gone home.

"Guess who was doing it? By herself? Me," said Beau. As she plugged away, she began to feel sorry for herself. And when it came time to vent, she found there was just one person she wanted to reach out to.

"I called Antje."

Antje did what a friend does. She commiserated. And then, shortly after they hung up, Antje showed up in person. She surveyed Beau's long face, the unpainted beige walls in the main meeting room, the lobby, and the upstairs bathroom. This was going to take all night.

As they began painting, the mood was grim. They vented to one another. Why hadn't the others planned to stay longer? But once the conversation began to yield to the sound of paintbrush on wall, Antje plugged her MP3 player into the speakers Andy had wired for the main meeting space.

There was Nelly. There was Grandmaster Flash. The monotonous motions of the work began to acquire a certain beat, slapping in rhythm to eighties and nineties hip-hop icons. Salt-N-Pepa. Notorious B.I.G. They began muttering lyrics under their breaths. Toes tapped, and hips twitched.

Run-DMC. Beastie Boys.

There was no one around to bother. The volume just got louder and louder. The work-like atmosphere drained away, and instead, it became fun. They laughed, and they danced.

By the time Coolio entered the proceedings, Beau's sadness was mirth.

"As I walk through the valley of the shadow of death," Beau shouted, dancing with a fluid grace that surprised Antje, "I take a look at my life and realize there's nothin' left."

They had chosen a blend of a peach-like color and burnt sienna. "So it was like, very warm. Not dark. Bright, but not overpowering. It was a lovely color," said Antje.

This might have happened during the painting session—Antje and Beau can't say for sure—but it was around this time that Antje made a proposition.

"Hey," she said, between tracks. "I want to go on vacation. Want to go to Mexico?"

"Oh my god, yes," said Beau. "We're going."

By 5 p.m. on moving day, when the volunteers began to show up to help KRI lug boxes, bean bags, office furniture, and pillows, the frigid atmosphere had begun to milk the clouds overhead of a light drizzle of rain.

In an effort to make things easy on the volunteers, the KRI members moved the most difficult items. Antje took the psychomanteum curtains home, but brought the other pieces to the new Center, hoping to reconstruct it. Andy left the drywall and the stack of lumber behind. He had no immediate plans for them.

Though Mike was still in jail, "he hand-wrote me a letter every single day," said Beau. Her husband, Troy, came with a truck and a trailer, which they used to move Mike's desk and office setup. He would have a place in Stratham.

The KRI members were also the ones moving the spirits.

Each medium went into the building, and made contact with spirit guides and entities that had helped them there. When it was Val's turn, she used a combination of imagery and meditation to draw the energy of each one into herself. Then, Val got into her car with her daughter Juliana, and Andy's two cats. She drove to the new Center, nestling the spirits inside her core. When she got to her destination, she parked and stood outside the new building, silently addressing them.

This is where we'll be now.

And then, just like that, the spirits were gone. Only time would tell whether they would stay at the new Center, return to the old place, or disappear altogether.

When she went back to the Center, Val kept her head down, as she had been doing all day.

"I just did what I was told and tried not to be in the way," she said.

But she noticed that people were talking about Andy, who was not present.

"What the heck?" Val heard somebody say. "This isn't good."

Beau went to check on Andy. This made sense. For all of the criticisms Andy tended to spin out into the world, Beau alone remained untouched. They were like siblings, and they had come to rely on one another for various forms of validation.

"I talked to Andy about everything," said Beau. "I talked to him about all the ups and downs of my marriage for all that time. You know?"

When she came back, a little while later, one of the volunteers asked her, "Where's Andy?"

Beau made a vague excuse.

"You know, he's just using you," the person said to Beau.

"Well, you don't see everything he does behind the scenes," Beau retorted.

But behind her facade, Beau said she was feeling irritated. The first glimpse of a problem had come earlier that day, when Antje and Beau discovered that Andy had not packed up all of his belongings. They both remember Andy having an emotional crisis that locked him up and prevented him from progressing with the move.

"It's like stacks of things everywhere," Antje said. "It's like, the mad professor."

Val said that Andy's lack of preparation was a difference, not a shortcoming.

"That's who Antje was. That's not who Andy was," said Val. "He didn't move the way Antje would move. She's not the most compassionate person."

Andy said later that he did not recall any particular emotional distress on that day. But the others said that he wound up sitting in his car. When Beau joined him in the car, she said he was teary-eyed. She offered words of comfort and support, but they were at odds with her actual feelings.

I am completely taking care of you, Beau thought, *and you're not doing anything.*

Beau said she could have forgiven it, if Andy had been able to articulate a trauma that the move was causing to come to the surface. But all he could say was, "I can't go in there."

Val thought that Andy was having an entirely natural reaction to the loss of the Center.

"He really put a lot of blood, sweat, and tears into the actual place," she said. "That can always be hard to leave. It had to do with realizing a dream: I managed to do this but now we have to leave all of that behind."

At some point, as I asked the members of KRI about the move, I became aware of an idea that Andy had some personal effects that had to be packed and moved, and that something in those personal effects grossed out some of the volunteers.

Specifically, a condom.

Andy said he had no condoms, and that there were in fact no condoms in the Center. The objective fact of the condom, or its absence, has been lost to history; but the individual experience of the prophylactic cannot be denied.

Antje was apoplectic.

"I had to pack his fricking condoms for Christ's sakes," she said. "Oh my God. Gross. That is like, it's gross. Why did I have to do that? So I was, I was, I was royally mad."

All of the conflict over moving day was good for one person, and one person only. And that person wasn't even a person. How the parasite must have feasted that day! Or perhaps it was panicking and using every ounce of effort in a doomed attempt to keep Andy in the old Center.

The KRI members had to decide what to do with the entity. Keeping a spirit bound to an earthly location was, to Beau and the others, highly unethical. And in some places, it could have even been considered illegal.

In the so-called "Ghostbusters case," a New York appeals court in 1991 ruled that Helen Ackley, a Nyack homeowner, had erred by failing to disclose the presence of a poltergeist to a buyer, Jeffrey Stambovsky. "As a matter of law, the house is haunted," the appeals court wrote, as it released Stambovsky from the purchase contract.

But there were no such moral considerations for inhuman parasitic entities. And this is why, with great satisfaction, the KRI members decided to leave the parasite in the retail center.

Beau saw a certain poetic justice in leaving the entity behind to plague the buttoned-up property manager who had forced them out. She sent the manager a mental message.

Like, good luck with that one, dude. Thanks for being such a dick.

For Andy, the decision to leave the parasite behind was a bright spot. And, he noted happily, much later, "They didn't rent that space out again for years."

Free of the parasite's influence, the members of KRI would come together to make a fresh start.

Or, they wouldn't.

Twenty-Six

YOU FREAKING DICKHEAD

The place has about it an indescribable soothing atmosphere of respectability and comfort. Here rest the remains of the principal and loftiest in rank in their generation of the citizens of Portsmouth prior to the Revolution—stanch, royalty-loving governors, counselors, and secretaries of the Providence of New Hampshire, all snugly gathered under the motherly wing of the Church of England. It is almost impossible to walk anywhere without stepping on a governor. You grow haughty in spirit after a while, and scorn to tread on anything less than one of His Majesty's colonels or secretary under the Crown.
—Thomas Bailey Aldrich, *An Old Town by the Sea*

"We're going to talk to the ghosts the same way I'm talking to you. With decency and respect."

Lou Calcagni, an imposingly bulky man wearing a Patriots winter hat and a camouflage jacket, looked sternly at we thirty-or-so trembling would-be ghost hunters. "We expect you to do the same."

Just two weeks ago, someone who had not treated the ghosts with respect had been scratched by a spirit—actually wounded by furrows in the skin that did not, Lou emphasized, look to be of human origin.

We sat on steel chairs pulled up to steel tables, surrounded by steel walls, and were trembling not in fear of Lou (though that would be an understandable response), but with the cold of late November.

"This is a steel box in cold water," Lou said with, I thought, a hint of satisfaction at our suffering. We were in the belly of a warship nicknamed the "Sea Witch." I had come to see a well-known ghost-hunting group, the Greater Boston Paranormal Associates, in action. Lou, a navy veteran, had served on similar ships, and his respect for the dead servicemen made it clear he would not tolerate their abuse.

"If you're here to antagonize the ghosts," he continued, "just tell me and I'll escort you off the ship right now."

No one answered Lou's long pause, or challenging gaze.

An hour later, in the ship's galley, Lou had lost all sense of composure. "You freaking dickhead!" he shouted, swiveling his head toward spirit or spirits unseen. "You fucking clown!"

Lining the walls, we watched Lou's tirade mutely, mouths agape.

This was my introduction to the world of paratourism.

The fans of nearly every fringe scene, whether garage bands, mixed martial arts demonstrations, farmers markets, sex clubs, or combat robots, want desperately to see an influx of money that will allow things to be mainstreamed, upgraded, and professionalized. The instant this is achieved, they're free to reminisce about how great it was before money ruined things.

The believer community is no different. There are near-constant grumbles about the commercialized inauthenticity of paranormal reality shows, even as profiteers are hard at work commoditizing every conceivable facet of the field.

Before TV shows popularized ghost hunting, the money was negligible and concentrated on a few personalities, like the Warrens. But things have changed, and quickly, according to Amelia Childs Schwartzman.

In 2023, Amelia was splitting time between podcasting about New England's paranormal community and running a haunted vegan crepe restaurant in Wellesley, Massachusetts. (By which I mean the restaurant was haunted. The crepes were fine.)

"It's not just something stoners in high school would do, breaking into an abandoned building to try to find a ghost," she told me. "There's high end, specialized equipment. The financial cost can get really high. You get really into it."

She described a highly competitive ghost-hunting landscape in which everyone is trying to monetize their interest, or capitalize on someone else's. Authors, TikTok influencers, television producers, conventioneers, tour hosts, hoteliers, and barkeeps are all trying to build audiences.

"There's a lot of ego involved," Schwartzman said. "You want to be billed as a big name."

There's a glut of ghost-hunting equipment, mostly electronics that riff on the basic equipment used in the television shows. And if you want to find a gift for the ghost hunter who has everything, consider the Boo Buddy, an electronics-laden teddy bear designed to draw child ghosts into conversations. Equipped with sensors that detect temperature and EMF changes, it speaks in a childlike voice, firing an endless stream of questions into the air. "What's your favorite color?" it asks. "Can you count with me? Mmmm. I like cookies. What's your favorite food?" It costs $360.

Because paranormal interests are not a category of goods and services that people track, no one knows how much money is generated by book sales, ghost-hunting television shows, mediums, house-clearing services, or ghost-hunting gadgets. Some estimates suggest the whole scene is a multibillion-dollar enterprise.

The only subset of paranormal commerce that experts track is paratourism, which is valued at a hefty $300 million annually.

Experiential tourism has been a lifeline for all sorts of wonderful causes and hobbies. Under this banner, ecotourism props up conservation efforts, culinary and wine tourism push back against the globalization of food systems by celebrating aspects of the "slow food" movement, and volunteer tourists build social good projects in underprivileged communities.

It's hard to fault such efforts, which have tied their marketing to a direct social good. But what is the impact of experiential tourism's hottest trend?

Ghosts are unquestionably the new hotness, to an extent that can be downright ghoulish. A waiting list runs for people to pay $380 to visit Fall River, Massachusetts, where Lizzie Borden murdered her parents with an ax, so they can sit in her kitchen, eat the same breakfast (sans mutton) she had on the morning of the killings, ascend the same steps, and sleep in her room. Or, if they prefer, the room where her mother's body was discovered.

"The money behind ghost hunting is growing—it drives tourism and helps create a sense of place," Dr. Michele Hanks told me. Hanks is a New York University anthropologist who does ethnographic work with paranormal investigation groups. "Part of the reason it's so supported and embraced by the local economy is it fuels tourism and place making," she said.

Ghosts can be shockingly valuable assets. One of the first sites to demonstrate this was the Ohio State Reformatory, of Mansfield, which was built in 1886 and has been long designated by the federal government as a historically significant site. A court order closed the reformatory in 1990, due to inhumane and crowded conditions for its prisoners. But historic preservationists had no money to maintain the structure, and officials were reluctantly moving forward with a demolition plan. In a desperate move, preservationists turned the reformatory into what was, at the time, the only former prison in the country to offer ghost hunts. Thirty years later, it draws 130,000 visitors annually, many of whom come to experience one of several ghostly experiences on the menu, including a ghost-hunting class for kids as young as thirteen. The reformatory has defined local tourism, and other Mansfield attractions have followed suit to create a vibrant ghost-based regional economy. Even the government-run state tourism department promotes breathless reports of shadow figures and full-body apparitions at these sites.

One potential issue is that, in a vast unregulated landscape, any ghost-hunting yahoo can assert that a site is haunted or unhaunted, often with very high stakes hanging in the balance. For example, civic leaders in Waco, Texas, crafted an economic development plan that tried to breathe new life into a sagging neighborhood. The centerpiece of the plan was turning an old bottling plant into the Dr Pepper Museum. And a critical part of the museum's success relied on allowing visitors to interact with the spirits of bottlers past, perhaps including Dr Pepper creator Charles Alderton, a Waco pharmacist who died in 1943. That's a lot of weight to put on the shoulders of insubstantial spirits, but they don't seem to be complaining. To burnish the value of its offered thirty-dollar ghost-themed tours of the building, the museum was soon touting that it had been officially declared haunted by McLennan County Paranormal, a local group. But of course, there's no particular reason to have faith in that group's assessment. It's run by married couple Mike "Ghost" Jacobus and Cindy Jacobus. The Jacobuses—er, Jacobi?—have unusual ideas concerning spirits. They believe, for example, that Ascended Beings make a living human tingle, while the Earthbound Deceased leave people feeling short of breath. And there's no way to evaluate those claims, because there is no agreed-upon body of facts that would adjudicate the truth of such interpretations.

This is a natural outcome of deconstructing the body of knowledge that is curated by our public institutions. In an institution-dominated world, facts drive beliefs. In a post-institutional world, beliefs drive facts. This is particularly good news to profiteers, because beliefs are easier to manipulate than facts. When it comes to the Ohio State Reformatory or the Dr Pepper Museum, those selling the tickets will, absent accountability to facts, perpetuate belief at all costs.

Dr. Hanks said ghosts are a particularly good commodity, because they can be incorporated with little effort into existing business plans. She knows the owners of a British pub that bills itself as the "most haunted pub" in the region, though the owners behind the marketing don't believe in ghosts at all.

These sorts of dynamics are playing out all across the country. Every state in America has dozens of haunted locations that are eager to receive visitors, many promoted by the traditional cheerleaders of capitalism—chambers of commerce and tourism departments.

A 2020 study in *Cornell Hospitality Quarterly* found that the financial value of a sensationalistic ghost story sometimes comes into conflict with the true history of the very sites they are meant to preserve. One particularly troubling example of this can be found in the former slaver states of the American South, where financially struggling plantations increasingly rely on funds from ghost tours. In order to attract tourists, the plantations spin lurid tales that hinge on the misery of enslaved people; however, as MacArthur Fellow Tiya Miles found in researching her book *Tales from the Haunted South*, many of the stories told are either grossly inaccurate, or wholly made up, with the fictionalizations often portraying the slave owners far more sympathetically than they deserve. I don't know what reflects more poorly on the community—rooting endlessly in the pain of actual historical victims, or feeling a need to create imaginary suffering to fulfill a need. For thousands of locations in America, history is being literally rewritten to highlight the most lurid details in a way that maintains appeal in a competitive market. This is why, after I plunked down fifty-three dollars for a ticket to board the USS *Salem*, a 1940s-era warship built to house 1,800 navy sailors, I learned only a few bare historical sketches of the ship, including its spooky nickname of Sea Witch.

Lou Calcagni's presentation was focused on the roughly one hundred people who died on board during its ten years of service. He talked about the babies and mothers who had died in 1953, when the ship briefly served as a floating hospital in the wake of the Great Kefalonia Earthquake off the southern coast of Greece. I learned about suicides and fatal accidents, including the time someone double-loaded a giant artillery gun during a training exercise, causing it to explode. Though part of me was irritated at the ghoulish lens, I knew that I was at least partially to blame. If it hadn't been for a decently packaged ghost presentation, I would never have gone there in the first place.

The scale of the *Salem* is massive. Anyone approaching its seven-hundred-foot length is rendered absolutely insignificant by the steel-gray guns and cannons looming overhead. As I boarded in the night, I felt like a Doozer on a ship built for Gorgs.

On each of seven decks, there were two main hallways running from fore to aft, with the center of the ship occupied by rooms and short connector hallways. Some of the rooms along the ship's exterior, like the chaplain's office, were tiny. Others, like the mess hall on an underdeck's stern, were huge. Steel can wreak havoc on an electromagnetic field—it can shield or blunt some fields, and it can be magnetized to generate its own field. So I wasn't sure how GBPA's EMF detectors would work here, where everything, it seemed, was made of metal—heavy locking hatches that separated one chamber from the next, the bed frames, the freshwater holding tanks, the countertops in the mess hall, and the base for the antique-style black rotary phone mounted on the wall.

During a brief tour, one of the GBPA members ticked off a few of the paranormal events that had been experienced on the ship—heavy footfalls, shadowy forms, and even some inexplicable manifestations of mist that had wandered around like independent entities. The group thought there were two interplanar portals that sometimes opened to dump fresh spirits onto the ship.

My fellow tourists ranged from the secretly skeptical to the open-minded and curious, to the heavily attuned, which included a young woman who was talking to her boyfriend about a tightness of chest and other sensations she was experiencing in certain areas of the ship.

"I love this stuff, but at the same time, sometimes I hate what it does to me," she told him. He gave her little reaction. It was her chest, she reminded him.

"I love this stuff, but at the same time, sometimes I hate what it does to me," she said again. I struck up a conversation with her. She was twenty-two, and had a nine-year running relationship with a child ghost that was perpetually five years old. She told me that her boyfriend was fine with her spiritual sensitivity.

"He knew that I could see dead people when we started going out, and he said it didn't matter." She explained this while looking at me, but with her mouth approximately six inches from his ear. Then she turned to face him. "You knew I could see dead people when we started going out, and you said it didn't matter."

Mercifully, the GBPA members took that moment to announce that it was time for them to initiate contact with the spirits of the *Salem*'s former sailors. They herded us through darkened corridors toward the rear of the ship. I had expected to hear squeals of a vessel being tested by the immense weight and pressure of the ocean, but the only sounds were our footsteps, traipsing single file along the metal floors, stepping over the bottom rims of the doors. We entered a cavernous mess hall and sat on metal stools facing inward, with our backs to rows of metal counters.

GBPA cofounder Don DeCristofaro, an old-school Bostonian, supervised as Lou and the others set up their equipment. Don wore a navy-blue sweatshirt and had a big bushy white beard that obscured his face. He walked with a limp due to a bad knee. He had grown up just a couple of miles away, and felt the weight of a long-standing personal connection to the ship itself, which had been in the harbor for his entire life.

The gadget-heavy nature of ghost hunting was on full display. They set up about a dozen electronic devices in a rough row along the length of the room. There were K2 meters about the size of a crème brûlée. There were thermos-sized "Paralights" that also detected electromagnetism. There was a Static Dome, which measured static electricity and looked like a large black cupcake. There was also a cylindrical Rem Pod (or Radiating Electromagnetism Pod), with a black-and-red color scheme that made it look like Satanist-brand cottage cheese. I was getting hungry.

At least fifteen other instruments were also in use, some on the floor and some on the sidelines. The overall effect was that a gaudily lit Christmas tree had been murdered and its remains scattered on the floor.

Don and Lou stood in the center of the room. It was time.

"Is there anyone here with us?" asked Don.

There was a pregnant pause. Then, one of the K2 meters just a few feet away jolted into a spectacular light show. As it gave off a satisfyingly loud tone, it cycled through a quick pattern of red, blue, and green, indicating electromagnetic frequencies. The whole basis of the modern ghost-hunting canon is that ghosts somehow set off the meters by altering these fields.

"Is that you, Chief?" asked Don. Again, the meter responded with a light show.

Beeeeeyeeeep.

It was, for a new believer, an electric (or electromagnetic) moment; two people, one alive and one dead, exchanging information from separate planes of existence.

I was watching closely. Skeptics say that ghost hunters unknowingly trigger meter reactions by carrying recorders, cameras, and cellphones within range. Don and Lou did move about the space while they talked, but it was hard to match up their movements to the tones being generated by the meters.

As Don spoke to the spirit of Chief, it quickly became clear that GBPA members had established a fairly dense mythology surrounding the *Salem*. Over ten years of investigations, they had used the ship's old navy yearbook to identify individual spirits—the stalwart Chief, the mischievous Skip, and the reliable Allison. With each investigation, GBPA extended what was essentially a running narrative, like a soap opera. Tonight, there was an effort to find out who had scratched Zach, a young paratourist who had come aboard for an investigation only to suffer the wounds that Lou had warned us about. Everyone suspected a ghost named Shackleford, who was mischievous to the point of malevolence.

Answers were complicated because, they believed, some spirits enjoyed gumming up the works by pretending to be other spirits. Don tried to protect against this by asking screening questions that he knew to be false.

"You were a gunner, right?" he asked Chief. No answer.

"You were an engineer?"

Beeeeyeeeep.

Chief had passed the test.

Was I simply reading a pattern into random beeps? Due to the design of the meter, the tone always lasted exactly ten seconds, and almost all came from the single meter that was closest to Don and Lou. But the silences in between could last almost any amount of time—two seconds, five, twenty or more. Occasionally they went off when they weren't being addressed, but they almost never went off in the middle of a question.

Over about fifteen minutes of questioning, it came out that Chief, not Shackleford, had scratched Zach. Chief communicated, through beeps and silences, that Zach had done something provocative during a previous visit to the ship.

And then, before they could figure out what, the K2 meter went silent. Chief had disappeared, leaving a tantalizing dangling cliffhanger for a future episode of questioning. Next up was a spirit that Lou clearly felt some amount of brotherhood with. Without explaining who the ghost was, he addressed the spirit as "Laz."

"What's going on, chum?" Lou said. "How you doing?"

Beeeyeeeeeeep.

Pat chimed in. "Hey, Laz," he said. "Did you say, last time we were in here, 'I'm gonna fly as a civilian'?"

Beeeyeeeeeeep.

"Is that what you heard?" Lou asked Pat. "I don't know how you hear that, Pat, because, because, I must be deaf. I would love to hear his voice because I would love to know that it is him."

"So, Laz," Don said, "you're here right now."

Beeeyeeeeeeep.

"Laz," asked Lou, "are you with your daughter, Angie?"

Beeeyeeeeeeep.

Lou's expression went dark.

"That's not you," he said. Then, angrily, "You freaking dickhead."

It was a screening question. The actual Laz had committed suicide after his daughter died of cancer. The daughter's name was not Angie. It was Maggie. And the spirit had failed the test.

"Shackleford?" ventured Don.

Beeeyeeeeeeep.

"Asshole," Lou was simultaneously raging, and trying to explain his rage. "I just want to find out why he's such an ass. I want to find out why he does that."

"He's a trickster," said Don, with a verbal shrug.

A minute later, Lou issued a threat.

"I think what we need to do is get someone on board and force him to cross over," Lou said.

Beeeyeeeeeeep.

"Is that what you want?" Lou asked, incredulously.

Beeeyeeeeeeep.

Shackleford repeatedly confirmed that he wanted to cross over. The GBPA members seemed to think this was dramatically out of character. They acted like parents, feeling the need to follow through on a toddler's threatened discipline.

"Okay," Don said, reluctantly. "We'll do it. We can bring somebody aboard who can cross you over. We can do that. And that's what you want."

Beeeyeeeeeeep.

"Alright, brother," said Don, roughly. "Done."

Beeeyeeeeeeep.

He tried to pull it back, one last time.

"You're screwing with us, aren't you?" he said.

There was no answer. The meter went silent.

Soon after, Don called an end to the session.

As I watched all this, trying to evaluate, I was convinced that GBPA was acting in good faith. Things were too disorganized and too spontaneous to be carefully choreographed. And every GBPA member seemed too wrapped up in their own moment-to-moment agendas to be faking anything. If they were playacting, it was world class.

But were they simply fooling themselves? Were the EMF meters activated by a ghost acting in response to a question, or were they simply random occurrences? If the meters were triggered by devices carried by the GBPA members, it was even possible that they had unconsciously trained themselves to elicit responses. It reminded me of a story about a classroom of college students conspiring against a professor who liked to wander about the front of the lecture hall while he taught. The students paid him attention only when he stood in one particular spot. After several weeks, it was as if they had nailed him to the floor. But the professor had no knowledge of his own behavior change. Perhaps the meters were similarly training Lou and Don.

This was the first of what was meant to be multiple conversations with spirits, some with the whole group and some in smaller breakout groups. But, not long after the scene in the mess hall, the K2 meters all stopped lighting up. One session led by Don, and another led by Lou, both yielded no results.

After a promising beginning, the night had turned boring, and the organizers seemed both puzzled and worried about failing to provide a top-tier experience. We all trudged back into the main room, literally and metaphorically dispirited.

"I'm just mad because I don't know what the heck happened," Lou kept repeating. They tried to piece together a narrative that tied the abrupt departure of a spirit named Allison to the cessation of all activity on the ship. But nothing really resonated.

Don tried halfheartedly to stimulate a bit of group discussion, but it was nearly midnight. Everyone was stone-cold and bone-weary. After just a bit of polite back-and-forth, we all headed back to the deck, to disembark from the *Salem*, and get into our vehicles.

As I left, blasting the heat, my mind was awhirl with the experience. Though the night had been disappointing to veterans of the scene, I had found those thirty minutes in the mess hall to be energizing. Had I seen a back-and-forth conversation between the living and the dead? Or simply a group of people venerating a random set of beeps? My main takeaway was how invested Don, Lou, and the others were in providing us with a robust experience. Giving us reliable deliverables was the keystone of their operation, and simple good customer service. They had become dependent on the dead to provide a living for the living.

But what would they do if the beeps of their K2 meters never returned?

Twenty-Seven

SHE JUST STARTED DOING STUFF WITHOUT ASKING

And to men like him, I said, when perfected by years and education, and to these only you will entrust the State.

—Plato, *The Republic*

In the late afternoon sun, sweat trickled down Val's neck. The higher she climbed, the steeper it got.

It was 2015. Ten years on, the *Ghost Hunters* television show had spawned a revolution in reality television; by one count, 109 paranormal reality shows were produced and aired, with 109 producers applying their creative talents to convince the public that ghosts, or Bigfoot, or aliens, were real. Financially, it was a bonanza—so much so that the Travel Channel abandoned topics like food, beaches, hotels, and culture in favor of an almost all-paranormal slate of programs that resulted in a 15 percent bump in viewership.

KRI was not at the feeding trough. They were caught in a delicate, unremunerative straddle in which they were not scientific enough to attract support from the scientific community, but too scientific to engage in a baseless cash grab.

June had been a battle of wet against heat; now, in July, the sun was taking an extended victory lap, sending throngs of gardeners out to triage wilting lettuces and listless peas. In the short time between Val's childhood and her adulthood, New England's summers had become hotter, and its winters shorter. This was slowly tanking the state's winter recreational tourism economy. We know the financial toll climate change will take thanks in part to UNH doctoral candidates Elizabeth Burakowski and Matthew Magnusson, whose research found that each poor snow season costs the nation's ski industry $1 billion, and twenty-seven thousand jobs. These trends make the state ever-more-reliant on dollars generated by ghost-seeking paratourists.

Val thought the upper reaches of Doublehead Mountain, located north of the Seacoast, would be a refuge from the heat, but so far, that wasn't panning out. Val took it in stride. The mountain's northern peak was just an external representation, one in this material world that was neither the only, nor the most important, plane of existence to her. She had, in the fifteen or so months since the new Center opened in Stratham, climbed more challenging mountains within herself.

Val had begun hosting a monthly Empath Support Group. Leading a group was a major personal step for her. At first, barely anyone came, but slowly, a small group of empaths began showing up, many needing guidance from someone with Val's experience. She wasn't at Beau's level. Not yet. But it felt like the start of something.

This was Val's first camping trip since childhood. So she was grateful that the small group of friends laboring up the mountain beside her included a woodsman with plenty of experience. It was Mike Stevens, having paid his debt to society and been released from the county jail. He was working for Beau's husband, tiling floors. Sometimes Beau gave him a ride to work, but he often biked. Ascending over the New England woods he loved, he seemed happy to be free. Between the prison diet and the exercise, he'd lost more than a hundred pounds. He was dating. He looked healthier.

KRI was healthier, too—when Mike returned to the helm at Social Saucers, attendance had increased. It was funny. The paranormal community was full of aspirants trying desperately to attract social media followers with outlandish stunts. Mike didn't truck with that sort of thing. Instead, he'd gained a following organically, through his earnest efforts to help experiencer after experiencer. He had stuck to his mantra of "people, not proof." Maybe the supposed abductees he spoke with were kidnapped by aliens. Maybe not. That didn't matter to Mike.

"*Something* happened," he would say. "You're affected."

The depth of the trauma was underscored all the time. He had once helped a man who, after washing the night's dinner dishes, put a crockpot lid on the kitchen counter to dry. When the overhead light hit the little dome in a particular way, he'd had a panic attack.

"Grown men, who would knock your teeth out in any other situation, become just—yeah, it really affects people down to the core," said Mike.

One of the things that moved Mike was how difficult it was for Social Saucers attendees to talk about their experiences in other settings. Many of them were afraid to talk even with their own spouses or children, afraid that they wouldn't be believed. Others had told family members, and were still not believed. They were victims without a safe space.

His personal friendships and contacts had allowed him to build a national network of like-minded people, who stood ready to offer face-to-face peer support to anyone who needed it. That hyper-specific service of helping abductees slotted into a landscape in which so many services were still provided by UNH, which remained one of New Hampshire's most critical institutions. It provided the public with education, career opportunities, free legal advice, evidence-based knowledge of climate change, employment, trustworthy public news, and agricultural advice. UNH had even helped maintain, among other recreational trails, the very path that Val and Mike now labored up. Andy was there, too, huffing and puffing right along with them.

Finally, in the early evening, they crested and found a wooded pocket of land surrounding an old cabin with a corrugated tin roof, log walls, and an eighty-two-year-old stone chimney that looked like the end stage of a Jenga game.

The cabin was yet another reminder of an age of institutionalism gone dark. It was built by the Civilian Conservation Corps, which in the 1930s became the most popular program of Roosevelt's New Deal. It sent three million young men with few prospects to work in the woods in exchange for a meager paycheck. The rugged experience was meant to improve their morale, their character, and, by extension, the country. This particular cabin had been built by and housed some of the hundreds of young men who picked up the pieces after a devastating 1938 hurricane killed two billion trees and thirteen New Hampshirites.

That's all history now. Val, Mike, and Andy had rented the cabin for about forty dollars.

With hours to go before sunset, they built a fire, ate burgers and trail mix, and drank water to rehydrate themselves. A brown rabbit—"It was the size of a Buick," said Andy—entertained them before dashing away. It was, Val said, eerily silent. She'd expected crickets, but on this wind-ridden peak, there were few bugs to break the silence.

It was time for them to get to work.

They were hunting Bigfoot.

Antje and Beau had opted out of the hike, which was a relief for Val, whose empathic sensitivity to simmering tensions had been draining her. Moving away from the spectral parasite at the old Center had done little to dissipate the discord between Andy and Antje. If anything, rancor had increased.

"I can read the energy and sense the tension, even if everyone is trying to play nice," Val said.

Andy's now-blatant lack of respect for Antje's ideas was driving Antje to, on at least one occasion, literal tears.

She often vented to Mike, who responded with gentle wisdom. She shouldn't take Andy's behavior personally, because it didn't reflect on her, Mike told her. Andy was just being himself. That's how he was.

"He was always able to help me come back to a place of center," said Antje.

Antje had also been centered by her ten-day trip with Beau to the eastern Yucatán Peninsula. While there, they saw prominent Mayan temples, and also little-known ancient stoneworks jutting out of the earth

She Just Started Doing Stuff Without Asking

in the jungle. They visited local villages where they ate in the homes of grandmothers. They marveled at cenotes, natural water-filled cavities in limestone that were supposedly used by the Mayans for ritual sacrifice. When they got back, they did a joint presentation about their spiritual experiences in the region.

Andy was sour. He was used to exchanging snarky comments about Antje with Beau. But now, when he vented about Antje, Beau would awkwardly direct the topic elsewhere, or even praise her.

"I never thought Antje was a good medium but Beau is all of a sudden saying how great she is. After Yucatan," said Andy. "It's like no, something changed here."

Beau and Antje said that their bond long predated the Yucatan trip. As her friendship with Antje kindled, her friendship with Andy dwindled. This was just the first time Andy had noticed it.

Beau soon had an impactful conversation with her spirit guide, Walking Elk. He told her it was time for the next stage in her role of helping spirits.

"I need you to take everything that you learned, and condense it," he said. "Teach it to people."

Beau complied. She developed a new seven-week course that covered everything she knew about intuitive development. She wanted a blueprint that would avoid all the common pitfalls of mediumship and intuition, and allow people to hear the voices of the spirit world. She called it SAGE, an acronym for Spirituality, Alignment, Growth, and Empowerment.

But this project was also different in another way. SAGE, Beau told Andy, was to be her project, and hers alone. Andy didn't argue. He understood that she needed some distance, but her seriousness came into sharper focus when he offered to edit the curriculum and materials, as he had her book. He assumed Beau would welcome the help, but she wouldn't let him even read it. Instead, Antje was giving her a hand with it.

"All of a sudden she just started doing stuff without asking," said Andy. "I mean, before she would ask. Everything she did, she'd ask for my input. And she just stopped doing that."

When Beau unveiled the SAGE Method as a trial run to a group of her existing clients, she was happy with the results. She had, she believed, distilled the secrets of hearing spirits into a program anyone could follow.

"They came out of it getting really cool hits intuitively right outta the gate," she said, excited. "And it was like, holy crap, this is fun."

And Antje was replacing Andy as Beau's primary confidant. Antje was building her own clientele, doing energy work and readings. She was working toward becoming a certified SAGE instructor, and was student teaching and leading workshops on Beau's behalf.

But for Andy, Val, and Mike on the mountain, the KRI drama seemed far away. They lazed about, feeling their muscles tighten as they watched the sun slide down the darkening sky. As soon as it disappeared beneath the horizon, Mike cued up some audio, and they began to broadcast into the darkness.

Huuuuoraaaaaawr!

Val and Andy were startled by the sonic equivalent of a turducken—a lion's roar stuffed inside a chimpanzee's screech stuffed inside the eerie ululation of a loon. It was followed by more unearthly screams, some high-pitched enough to split eardrums and others so gruntingly guttural they rattled ribcages.

They were recordings of reported Bigfoot calls from around the country. In them, Val heard not just mindless noise, but emotions. One, she later wrote, had "a tone of desperation and despair."

They knew that this Bigfoot experiment was unlikely to yield a Solid Phantom, but just imagine! The White Mountains, and KRI, would become true epicenters of the national paranormal community.

Of course, the responding cryptid might not be a Bigfoot at all.

"Our version of a Bigfoot here in New Hampshire is called a Woods Devil," Mike later said, during an interview. "They're about what you would get for Bigfoot reports as far as height. They were a little more slender, more silver, and it was said they could turn sideways and you wouldn't see them. They were almost invisible."

He said that reports began in the 1700s, when the region was first being clear-cut for lumber, and continued up until 2003. "Maybe we still have them. I'm not sure."

They played the audio, then listened to the still night. Nothing. They played it again, and listened again. Nothing.

And again.

A 2022 YouGov poll showed nearly half of Americans, 48 percent, are open to the idea that Bigfoot exists, which fuels an ever-optimistic legion of cryptozoologists. Beyond Bigfoot, cryptozoologists scour the world for proof of various other beasts of legend. There are tales told of many such creatures.

In remote northern Myanmar there were stories of a monkey with an upside-down nose that, with every rain, had to sneeze to clear out its nose puddles. Others, along Africa's coastline, said there were giant shark-eating rays, bigger than a man, that could shock a person to death. And on Madagascar, there were tales of vampiric ants that sucked the blood of their prey.

But you may be surprised to learn that these three species actually exist. They were all documented. In 2015.

And 2015 wasn't unusual. The truth is, every year, humans identify bizarre new critters from among the estimated five to thirty million as-yet-undocumented species living on Earth. But these discoveries aren't coming from cryptozoologists. They're being made by institutionally affiliated zoologists and biologists. Historically, Team Scientist has outperformed Team Cryptid, by a score of roughly two million identified species to zero.

This is kind of remarkable, given that cryptozoologists and field biologists employ the same basic method, which is typically to simply search the habitat in which an undocumented creature supposedly lives.

But of course, the cryptozoologists have a much narrower focus. The term was coined around the 1960s by Belgian-French zoologist Bernard Heuvelmans and Scottish biologist Ivan Terence Sanderson, who spent much of their time arguing for the existence of bats bigger than an eagle, living dinosaurs, sea serpents, fifteen-foot-tall penguins, and, of course, primates with human-level intelligence like Bigfoot.

These are, essentially, monsters of the sort that populate folklore. What ties these creatures together is that they have been rejected as implausible by the scientific establishment. There is a key difference between the camps. Scientists try to document the creatures that the evidence suggests are most likely present. Cryptozoologists try to document the creatures that are least likely present.

Perhaps in an effort to explain their abysmal track record, cryptozoologist circles are rife with the idea that scientific institutions are ignoring or suppressing evidence of cryptids. Their antagonistic relationship with institutional science has made cryptozoologists a strange bedfellow: creationists, who hope that the discovery of a living plesiosaur will undo one hundred years of evolutionary theory in one fell swoop.

As with evolution and death, scientists have been hard-pressed to create a spiritually meaningful narrative about cataloguing the vast variety of life on Earth, an all-cattle, no-hat approach that has allowed cryptozoologists to flourish. Creationists, who have all hat and no cattle, are waging a bet that cryptozoologists will walk a reptilian steer into their paddock. References to cryptozoology figure prominently in Christian-lensed textbooks and creationist museums eager to prove that the monstrous behemoths, leviathans, and dragons of the Bible were in fact actual creatures placed on Earth to demonstrate God's might.

The same month that Mike led KRI to seek Bigfoot on Doublehead Mountain, creationist Dave Woetzel wrote an essay for *Creation Matters* magazine that identified the documentation of a Loch Ness–type creature as a critical goal for those who believe in intelligent design.

"I believe that the discovery of a living dinosaurian creature, in one of these remote regions, would provide a forum for the truth of creation, hasten the eventual demise of evolution, and open doors for evangelism," he wrote.

This line of thinking drove creationists to pour hundreds of thousands of dollars into cryptid-hunting jungle expeditions, and to use creationist organizations to promote and report on their findings. But there is a shortsightedness in this strategy. The church has been in a fistfight with science since 1633, when the Roman Inquisition jailed Galileo for the crime of describing Earth's orbit around the sun. But church and science increasingly resemble lobsters duking it out in a grocery store tank, totally unaware that spiritualists are about to eat them both for lunch. Rather than becoming allied frenemies, creationists have thrown themselves into a short-term strategy of bolstering the efforts of a paranormal community that, writ large, has no love of the church or its teachings. Even now, the persistent failure to validate Bigfoot's existence has the cryptozoology community considering alternative explanations that are in direct conflict with the Bible. For example, Val, along with many of her peers, believes that Bigfoot's elusiveness might indicate that it's an extraterrestrial or interdimensional being that can pop in and out of the woods at will, which is hardly in keeping with a biblical worldview.

I do have a suggestion for those biologists who seek to add to the storehouse of biological knowledge and the cryptozoologists who seek to upend that storehouse.

Field naturalists sometimes display a capacity for humor when christening a new species, which is why a Costa Rican bicep-laden beetle is named *Agra schwarzeneggeri* after Arnold Schwarzenegger. There is also a genus of Australian orb-weaving spiders named *Pinkfloydia*, an American black-helmed slime beetle named after Darth Vader (*A. vaderi*), an Ecuadorian tree frog named *H. princecharlesi*, and *S. beyonceae*, a golden-haired Australian horsefly named after Queen Bey.

Why shouldn't some enterprising biologists give us the Bigfoot Beetle, the Nessie Fish, and the Chupacabra Horsefly?

If they did, science and cryptozoology could at last agree that yes, Bigfoot really does exist. Though, ironically, the credit for discovery would accrue, once again, to Team Scientist.

Back on Doublehead, the yawps and hoots and shrieks continued to reverberate through the night. The KRI members didn't get the answer they were expecting. But there was an answer.

Though there was no audible response, Val said she noticed an almost palpable change in the air, something that she said was not simply a case of the heebie-jeebies. Mike and Andy felt it too.

Though they had come seeking adventure, sitting in the dark on a mountaintop suddenly seemed foolhardy. Had it been hot? Now, they were shivering. Adrenaline surging, they scrambled inside the cabin and bolted the doors. A two-tiered bunk bed stood in each corner. They sat on wooden benches and speculated about the cause of the feeling they had suddenly shared. They laughed at their own unease. Was this the reaction of brave ghost hunters? But they didn't resume the audio. For two hours, neither Andy nor Mike went outside to smoke, which may have been a record for both of them. It was late when the feeling dissipated and they finally ventured back outside like nervous deer, clustered near the safety of the cabin, to admire the starry sky.

When they took to their bunks, sleep came slowly to Val. She lay on her pancake-thin mattress, eyes closed, thinking about the experience. Her empath powers told her that they were being watched. Was it a physical being? A spirit? A trick of the mind? She mulled it over. Then—

Thunk!

Something banged, once, on the tin roof. If it was an acorn, it was an acorn the size of a golden retriever. Val opened her eyes, heart pumping, but there was no other sound. No one else was awake. She lay still, trying to find the narrative that would satisfy both the logical part of her brain and the gut feelings that she had learned to trust.

Something was out there, she concluded. After banging on the roof, it had retreated to observe from a distance. Once she sensed that it had no ill intent, her hike-weary body demanded precedent, and pushed her down into the darkness of sleep.

Twenty-Eight

FOR MY HIGHEST GOOD

Often, from childhood upward, they had seen it shining like a distant star. And now that star was throwing its intensest lustre on their hearts. They seemed changed to one another's eyes.
—Nathaniel Hawthorne, "The Great Carbuncle, a Mystery of the White Mountains"

The next month, Mike and Val made another hike, this time up the western slope of Black Mountain in Jackson, toward another uninsulated log cabin built in the 1930s by the Civilian Conservation Corps.

Everyone in the small group was panting except Mike's daughter, who picked her way lightly among the boulders. Val's empathic powers were operating in overdrive—everywhere, it seemed, there were signs of a spirit energy watching them. She saw a tall, slender nature spirit gliding up alongside the path. Patterns jumped out at her from the landscape—an alien face in the bark of a tree, an alien face in the leaves on the ground, an intersection of branches in the sky that formed a distinct alien face.

The landscape's alien theme made perfect sense to Val, who knew these images were as intuitive as they were visual. Today, they hoped to make contact with extraterrestrial beings from atop the mountain.

The idea had sprung from KRI meditations, during which Val and others had encountered spiritual beings that identified themselves as coming from somewhere other than Earth. Val wanted to take this further, and establish a spiritual connection to a physical spacecraft from another planet. It would be like operating a metaphysical shortwave radio from the ground to a UFO.

Some paranormalists have pursued these sorts of interactions, known as Close Encounters of the Fifth Kind. But some of the same pious leaders who want to absorb extraterrestrials into the biblical canon find this flavor of contact to be against the will of God. Tennessee representative Tim Burchett, who has said that his Baptist faith is only strengthened by the thought of aliens, told NewsNation's large audience that multiple members of the United States Congress have characterized CE-5 as an evil.

"People are opening themselves to demonic suggestions," Burchett said. "I know that people are doing séances to bring these craft down or in, or something like that, and to me it reeks of a demonic suggestion." He seemed to be motivated by worry that aliens could be understood in a spiritualist context, rather than a Christian one. "They talk about their spiritualism and things like that, and it has all the essence of a classic séance, and I worry about that," he said. "I really do."

Burchett would surely have frowned upon Val and Mike, who had spent the spring teaching a five-week KRI course, at a cost of twenty dollars per session, about "CE-5 Protocols." They'd spun off a monthly private workgroup. For four months, Val and Mike and the others had sat in a contact circle, sending their earnest intentions into outer space. In just those four sessions, the small group experienced, according to Val, "missing time, concurrent images in meditations, [and] unusual symbology presented to one or more of us." Val had seen more than enough to convince her they were onto something big. It was time to, she said, "take it outside under the stars."

It was late afternoon by the time they arrived at the cabin on the summit. After admiring the view, they went to a local natural spring to collect some water. Mike began clearing out debris to allow the water to flow more easily, while Val sat on a rock.

She heard music coming from the direction of the cabin. It was too faint and indistinct to make out in any detail, but when she pointed it out to the others, it stopped before they heard it. When she stopped talking about it, the music resumed.

Mike got a fire blazing and cooked up some Hamburger Helper, which they wolfed down.

As the darkness grew, Val cleared the ground of any spiritual impurities, and created an area of protection. She then called in her spirit guides, a rotating cast that often included a cat and a pair of male-female guides who balanced one another out. They would help her interpret messages from beyond.

"I can trust the communication I'm getting is accurate and for my highest good," she said.

They sat and watched the slow-breaking night. Three or four times, they saw a small light that would flash a couple of times and go dark. Now they all heard the faint music, and it seemed to be coming from the direction of the spring. But as soon as they tried to pinpoint the source, it faded.

Even the most staid of scientists agree that the night sky is full of wonder and mystery. The partial moon that emerged, for example, has strange things happening in its ever-dark, unimaginably cold crater interiors. We know this thanks to UNH research scientist Andrew Jordan, who in 2014 published a paper showing that latent electric charges in the moon's soil create flashes of tiny, noiseless lightning that vaporize bits of soil; it is the fluffing of this process that is chiefly responsible for the powdery texture of the lunar surface.

And in 2017, UNH gamma ray astronomy professor Joseph Dwyer uncovered causes of another kind of unusual lightning—gamma-based "dark lightning," which bursts upward from Earth's clouds during thunderstorms, punching through the atmosphere and affecting the longevity of our satellite systems.

In 2020, yet another UNH astronomer, Dacheng Lin, found a Solid Phantom of his own, when he observed, for the first time, a medium-sized black hole in the Gal1 galaxy, located 740 million light years away. Before that, we had only observed small black holes, just a few times larger in mass than our sun, and monstrously large black holes, a billion times more massive than the sun.

Against this backdrop of an electrified moon, dark lightning, and a full range of black hole sizes, Val and Mike watched satellites and planets, shooting stars and airplanes.

Finally, the moon set below the mountains. They were getting tired, and the dark cabin beckoned. Mike asked Val to check with her spirit guides. Should they retire?

Not yet, said the guides. Something will happen soon.

A few minutes later, Val pointed.

"That doesn't look normal," she said.

In lower altitudes, the past month had done little to improve things at KRI.

Beau was beginning to take stock of her worth. She had an amazing talent that she was using to help others, both on Earth and beyond. But a lot of her time had been taken up with organizational duties that supported the entity of KRI. "Advertising," she said. "Speaking engagements. Organizing, showing up, setting up, tearing down. All that was done by me." And she was beginning to feel that her own success was not dependent on KRI's organizational health.

Andy was as preoccupied as ever, and both Beau and Antje were pulling back from their traditional duties. That left Val and Mike to shoulder an increasing amount of responsibility for the daily operations of KRI.

"No one really wanted to be there anymore," said Antje. "Val and Mike were doing the yeoman's work."

In any given week, Val might be writing copy and designing flyers to promote a drumming meditation, registering attendees for a demonstration of a new piece of ghost-hunting technology, posting social media notices for an event that paired off psychics and mediums for interdisciplinary readings, or collecting fees at the Energetic Wellness Sampler, a spinoff of the Psychic Sampler that focused on health and healing.

The old Val might have seen it as a lot of drudgery. But now, she saw the upside as having a finger in all of the goings-on. She was getting a free front-row seat to presentations that contained all sorts of valuable information.

"It allowed me to get a paranormal spiritual education without having to dip into my pocket," she said.

At the same time, Mike was setting up equipment to livestream events, decorating the space with streamers and balloons for celebrations, reconfiguring the space for meditations and presentations, and of course building cryptozoologist- and UFOlogist-themed events by bringing in guest speakers or hosting discussions himself.

Beau was in a surprising state of transition. Her marriage to Troy was on its last legs. In addition to SAGE, she had rented the entire downstairs from KRI, for $1,100, to start a web-page-design company.

Where Val's path of self-development had led her to be more positive, Beau's path led her to a place that was more openly negative. She no longer wanted to paste a smile on her face, come hell or high water.

"I said, you know what? I'm not masking for shit. I don't care," she said. She hired employees, and worried less about ensuring everyone was at ease with every word she spoke. There were moments, hours, days, when she cringed at her own assertiveness. But over a period of months, she latched onto a dawning realization. "Wow, I can be direct and honest and people still care about me and they still think I'm a good person," she said.

From Andy's perspective, Beau's launch of the Idea Garage was perplexing.

"She bought her secretary a computer screen that was bigger than my TV," he groused.

He respected Beau's talent. But how could her spiritual path lead her to start an internet business, at the expense of her spirit-related duties at KRI?

"She's taking the path of being the hollow bone between the spirit world and the people who need their information," he said. "Could she do that while running a, a computer business?"

"This could be my path too," Beau told him.

"It's not," Andy replied flatly.

With Andy and Beau starting to complain about each other, Antje and Andy continuing to complain about each other, and Val getting continued social distance from Antje and Beau, things felt worse for Val. She was doing so much of the work, but the decision-making all seemed to be happening in a loop to which she was not privy.

"I was feeling fairly left out," said Val. "It was a very volatile place at that point. It wasn't easy to see, but I felt it."

At the end of each frosty and dysfunctional staff meeting, Val left with the same feeling of unease. Every time she walked out of the Center into the sunshine, she would question whether she had a future there. It was wearying.

"I have to quit," she muttered to herself one day.

But then a voice inside her spoke up.

Hang on, it said. *A little longer.*

Back on the mountaintop, when Val showed the others the light she had spotted, it began to pulse.

"Hey, guys," said Val. "Do stars do that? Because I've never seen a star do that. Do planets do that?"

They knew that their guides had told them to stick around for a reason. Val watched it come up over the treetops, right in the center of her field of vision. She judged it to be bigger than a star, and it was throbbing and changing colors.

First white, then red, then green, then white again.

These colors are actually a common way to experience the stars Capella, Arcturus, and Sirius, and they are also commonly used by aircraft or drones. But then the light they were watching began to behave in ways that no star or aircraft could.

It moved up. Then down. Side to side, and then zig-zagging. It looked, Val thought, as if it was dancing. She was happy that, of all people, Mike was the one sitting by her side. What had begun as a thought exercise in the cloistered room of the Center led to this exhilarating, open-air sighting.

But Val sensed empathically that something was going on with Mike. She took her eyes off the light to look at him.

"When he realized, that's not a star, it was unlike anything I'd ever felt from somebody," Val said. "He was like, trembling."

To Val, the light was beautiful, winking in festive colors.

"But Mike's instinct was to be afraid. That was the only time I had ever seen him like that," said Val.

She leaned toward him.

"Are you okay?" she asked.

"Yeah," he replied, tentatively. "I think so."

But she knew he wasn't, not really. Val said that seeing him in emotional pain was sad, but she was eager to help him through the experience, as he had guided so many others. "I was able to be a friend and be understanding and just be there for him in that moment."

Inwardly, she consulted with her spiritual guides. Had they inadvertently connected with something that was not benevolent? Her advisers struck a reassuring tone. What they were seeing was not harmful, she was told. She relayed that to Mike and, over time, he seemed to slowly come back to himself. Finally, he was calm.

They focused their intentions on the craft, as a means of communicating with its distant pilots. Would it circle in the sky for them?

As they watched, gape-jawed, the light began to soar up, then down, describing a tear-drop shape in the sky. That's when they knew for sure that it was a full-blown CE-5 experience. After about forty-five minutes, a scrim of clouds glided into the sky, and the light faded away. They waited, but it never came back.

Val's spirit guides told her that the encounter had been designed to satisfy their own personal desire for knowledge, not to provide proof to the world. It was, explicitly, an experience, not evidence.

In the cabin, once Val was wrapped in her sleeping bag in bed, she heard tiny things scurrying for cover and looking for food.

Val was a light sleeper. She listened to the mice and waited for a sort of energetic equilibrium to happen. She was like a deep-sea diver. Her body needed to adjust to the energy of the environment before she could relax.

She had a bottom bunk in the front. After nearly an hour had passed, she heard Mike get out of his bed. He turned on the lantern. Without asking, she knew something was wrong.

The others were still asleep. Mike told her, in low tones, that he could hear something moving out back. When Val got out of bed, she put her ear close to the wall near Mike's bed. After a moment of silence, Val and Mike started violently. It was the sound of fur rubbing against wood. Something heavy, like a moose. Or a bear. Or something else.

The sound was slow and deliberate. Schhhhhhhsh. Pause. Schhhhhhhsh.

A small footpath ran along that side of the cabin. They heard twigs bending and snapping as it moved away from the cabin, toward where the path headed steeply uphill toward the mountaintop. Mike had a trail camera set up there.

But the following morning, the camera showed no trace of what they had heard. Moose, bear, Woods Devil, alien. In the extraordinary universe in which they lived, who knew?

Twenty-Nine

THAT'S NOT INTUITION

It is just as you say. But we won't talk of it. Of all ghosts the ghosts of our old lovers are the worst.
—Arthur Conan Doyle, *The Memoirs of Sherlock Holmes*

Antje sat in her small KRI office in the basement level. She heard a voice in her head.

Go. Now.

It was the same voice she had heard when she left California. Antje didn't hesitate.

I'm leaving, she told her guide, silently.

She walked into Beau's office, across from hers.

"I'm going," Antje told Beau, in a conspiratorial voice. "Right now."

Beau wasn't sure what Antje meant.

"I'm leaving," Antje said. "I'm going to tell Andy right now."

"Whoa, whoa, whoa," said Beau. The tension was suddenly electric. "What is it you're planning? What's going on?"

"Nope," said Antje. "I've got to go right now."

What struck Beau later was the *speed* with which Antje made the move. One minute, she was talking about upcoming events. The next, she was walking up the stairs toward Andy's office.

Antje had learned that she needed to act quickly on intuition.

"If I'm given too much time to stew about something, then it becomes all clustered," said Antje.

And besides, it was no secret that Antje was unhappy at KRI. Every time she entered the building, a sense of dread settled upon her.

"I didn't want to go in there," she said. "I felt gross going in there. It felt gross."

Antje had begun at KRI with high expectations.

"I can come in here and put structure around this because dammit I can do that," she had thought. "I'm really good at that. That's what I can offer. I have these skills that other people may not have."

But KRI had come to represent a personal failure. She thought she was an irresistible force, but she had met in Andy an immovable object. The result was a stalemate, and a stalemate was, for her, a loss.

The psychomanteum had never been rebuilt. After the bumbling trainee stepped on the nail, it turned out that he had not signed the release form she had so painstakingly crafted. She could create all the forms she wanted, but if KRI wasn't going to adopt them, risks would be forever parked on their doorstep. Antje had been forced to drive the trainee to the hospital and hand him $300 in cash to help cover the cost of the injury.

"You know what?" she found herself thinking. "Andy's going to do what he's going to do. I can't work with this."

When Antje told Andy that she was leaving KRI, there were no fireworks. He was even gracious. Hearing that she would no longer be agitating for various things to happen at KRI made him not the slightest bit sad.

Within minutes of telling Andy that she was leaving, Antje opened a web page of real estate listings. An office rental jumped out at her, in nearby Newmarket. She picked up the phone.

"An hour later, I was here, signing the lease," said Antje.

Privately, Andy thought Antje wouldn't last as a medium without the KRI brand to prop her up. And then, not so privately, he told Val and Beau and Mike that she wouldn't last as a medium without the KRI brand to prop her up.

A few days later, Antje came back to clean out her office. That's when the tension broke. The conversations between Andy and Antje quickly devolved into disagreements over a slew of material things that I, several years after the fact, found difficult to inventory, let alone resolve to my own satisfaction. It reminded me of a couple mired in divorce. It seemed to boil down to a question of Antje's acknowledged generosity. Had the various items she brought in for use been donations? Or loans? At issue were lamps, pillows, a mirror that wound up in Andy's closet, a painting that hung on the wall of the main meeting room, and an old film camera, a Canon, that had belonged to Antje's father before he died.

There were two competing, compelling narratives. In one, an embittered Antje lashes out at Andy by taking anything from the Center to which she had any sort of tenuous claim. In the other, a fit of pettiness seizes Andy, who takes many of her personal items that happen to be in the Center and locks them away from her.

Then they started to fight about money.

Antje had prepaid her rent through December. To Antje, the prepayment was intended as a favor, but now she wanted most of the unused amount—nearly $1,000—to be returned. To Andy, they had a verbal commitment to a sublet that went through the end of the year, and one that he was not obligated to release her from. But Antje distinctly remembered that when she bought the insurance, Andy had told her to "put it on my tab." Now, she said, his tab was due.

Andy eventually agreed to return the rent money, but not the insurance payment.

"It was like, 'What the fuck, take the money and get out of here,'" he said.

Val, again, felt out of the loop. Antje never had a significant parting conversation with her, and so she only heard Andy's side of things. In the end, Antje left with her mirror, her friendship with Beau, and plans to start her own spiritual venture.

As Antje and Andy battled it out, Beau was keenly focused on the nuance: What, exactly, was Antje saying? And how was Andy responding?

Her stake in it went beyond her personal relationship with the two of them.

Beau, too, was leaving KRI. She and Troy were divorcing. She wanted to pursue SAGE and her internet business from the Pacific Northwest. Maybe Seattle.

The seeds of her departure had been sown during the move from the old Center to the new one, when Andy, perhaps under the influence of a parasitic spirit, had broken down.

"By the end of that moving day, I think that was the beginning of the end," said Beau.

Antje had ripped the Band-Aid off, but Beau inflicted maximum discomfort on herself, peeling every millimeter of grimed adhesive from every now-graying hair on her body.

"The confrontation piece was terrifying for me," said Beau. "It wasn't for her."

She had framed it at first as a possibility, and then that possibility became increasingly likely, until she had fully eased into a new reality. It was time to move on.

To Val, losing Antje was losing a valuable asset. But losing Beau felt like losing the Center itself. It worried her.

"This person who brought so much to the Center is leaving," she said. "What are we going to do without her?" Val's relationship with Beau, never the strongest, was fading. Val made peace with the loss fairly quickly. "I hated to see Beau go, but knew she needed to do what was right for her."

Andy, on the other hand, felt strongly that Beau should stay at KRI. For the first time, their relationship was freighted with an existential tension. As a gulf threatened to divide them, parts of them clung together. Beau said she felt like a wounded little girl.

"I wanted to be able to pursue my life, and not lose my friend," she said.

Andy, for his part, didn't understand why, after years of partnership, Beau was holding him at arm's length. He didn't even know what SAGE's curriculum was, until he found a copy of some class materials floating around the Center.

"Typos," he said.

Clearly, Beau needed him.

That's Not Intuition

To Andy, all of Beau's behavior—the divorce, the shutting Andy out of SAGE, the internet business—led to a fresh conclusion. Beau, he determined, had strayed from her spiritual path. He suspected that she had lost both integrity and her ability to tap into the spirit realm. Andy told Beau that she was no longer an effective medium.

"She just completely refused to see any of it," Andy said.

He told her that he was using his intuitive powers to confirm that she had lost her intuitive powers.

"That's not intuition," Beau said. "You're not seeing that. That's not happening."

"You can't see it. You just can't see it," Andy told her. "You told me what your path looked like, and this ain't it. You're not seeing clearly. You're lost."

For years, Beau and Andy had crafted a world that balanced Andy's propensity for rational proof and Beau's embrace of individual experiences that were more meaningful than mere facts.

Now, they were both at loggerheads with their own belief systems. Andy could produce no scientific evidence of his intuition that Beau had lost her path and her contact with her spirit guides. And Beau could not accept his individual experience, because it contradicted her own.

This very same dynamic defines the antagonistic relationship between modern American institutions and spiritualists. When churches or medical science or the justice system come into contact with ghosts or aliens or parasitic entities, the scientific evidence upon which they rely demands that the individual experience be categorized as heresy, or psychosis, or irrational criminality. These framings so threaten the self-identity of the spiritualist that the only human response is to invalidate the institutions themselves. And if this happens with enough people, and with enough force, then the institutions are at risk of their own identities being threatened—in a great many cases, facts define reality only to the practical extent that people agree to acknowledge those facts. As Andy and Beau each refused to acknowledge the other's incompatible worldview, they both lost something.

Beau's stated departure date drew closer.

She and Andy began to bicker over things. Andy said that Beau's use of the big meeting space for SAGE classes was contingent on her meeting her membership duties.

"We had a verbal contract. She would provide X number of meetings she would personally supervise," said Andy.

When Beau withdrew from hosting the Lifting Lodge and Sacred Space events, Andy argued that she should pay usage fees for her SAGE classes.

"It really pissed me off that she stopped doing all of it and then started taking up all the time from the big room to teach SAGE, which was really lucrative, without giving me any cash," said Andy. "It's like, wait a minute, you, you had a contract here. I don't want to call it a confrontation, but I did bring that up."

"Actually I can just do that in my office," Beau said. And she did.

Beau said that Andy convinced some of her clients that she was no longer psychic.

"People started coming to me and saying it was so sad that someone with my abilities would have lost them. I'm like, 'What the fuck, dude?'"

Over the years, Andy had become friends with Beau's husband, Troy, and regularly played cards with him. One night, Beau said, she came to Andy, looking for a sympathetic ear about her divorce. But instead of his typical sympathetic ear, she said, she heard Troy's arguments coming out of Andy's mouth.

"I was stunned," said Beau. "I was more hurt in that moment. . . . I was like, 'Andy, you and I are best friends.'"

Andy said that the make-or-break moment came for him when he told her that, over a game of cards with Troy, Andy learned that Troy didn't support her Idea Garage venture. Before he told Beau this, Andy swore her to secrecy, but she immediately went to Troy with the accusation; Andy said that the betrayal of his confidence showed him that the friendship was dead.

At various times, I tried to get Andy to talk about how he was feeling at the loss of his friend, Beau. He consistently parried.

"Did that hurt for you?" I asked Andy one time.

"You know, uh," he answered. "Antje and Beau just verbally renewed their leases. Uh, they changed offices. Antje took Beau's upstairs office.

Beau took Antje's downstairs office, and, uh, literally they, they prepaid three months and wanted it all back. It was that quick. . . . I mean, I literally was in the process of repainting, uh, Beau's former office. . . . But at that point, the relationship had already deteriorated to the point where, I don't even want Beau upstairs anymore."

He only came close to expressing his pain once, when he said that "her demonstrated lack of integrity was pretty devastating."

When I told Andy that Beau maintained she had not lost her spiritual path, he, in his typical way, went through an on-the-spot, thinking-out-loud rational assessment of his own assertions in which he strove for objectivity.

"I don't know if I'm the only person who noticed or maybe it's some internal construct where, you know, I was just not feeling it anymore," said Andy. "I don't know what happened, but I, to me, and the, the evidence of what she said and the kind of things she was talking about supported the structure that she can't do this anymore."

It was a valiant attempt at a scientific answer. But the question may have been an intuitive one.

Finally, Beau's last day arrived. Her final event on the KRI calendar was a SAGE course, at a cost of $1,500, with Antje as a co-instructor. Rather than using the KRI space, they offered it as a virtual class.

Beau, like Antje, never had a direct discussion with Val about leaving. Val said she was a little hurt, but not like Andy.

"He was so close with Beau," Val said, eight years later. "He's still wounded to this day. It's a fresh wound. It hasn't scabbed over."

Thirty
CONSIDER THE JINN

> *We early found ourselves spending many hours in efforts to secure support for deserted women, insurance for bewildered widows, damages for injured operators, furniture from the clutches of the installment store. The Settlement is valuable as an information and interpretation bureau. It constantly acts between the various institutions of the city and the people for whose benefit these institutions were erected. The hospitals, the county agencies, and State asylums are often but vague rumors to the people who need them most. Another function of the Settlement to its neighborhood resembles that of the big brother whose mere presence on the playground protects the little one from bullies.*
> —Jane Addams, *Twenty Years at Hull House*

The triumph of paranormalism over institutionalism in New Hampshire is far from complete, but the signposts are frightening, the trends undeniable. Even when the economy roars, institutions keep whimpering.

In the early 2010s, some 96 percent of Democrats and 99 percent of Republicans expected their children to go to college. In 2021, just ten years later, 46 percent of American parents said they preferred that their children *not* attend a four-year college.

The toll of this distrust has been steep. By 2023, more than a dozen of the state's colleges had closed, along with twenty-six labor and delivery units and thirty-two churches.

And some of the institutions that remain are shadows of their former selves. UNH's doors remained open, but its fortunes reminded me once again of the sorites paradox—as financial pressures caused administrators to eject component after component, at which point would it no longer constitute a heap of college?

With public funding in the toilet, and enrollment dropping annually, in 2017, UNH told seventeen lecturers that their contracts would not be renewed. Over the next few years, UNH cut more lecturers. It cut its Japanese program. It cut programs in civil technology, horticultural tech, integrated agriculture management, and culinary arts. The programs most at risk were often ones that were begun with federal grants in the hopes that their worthy goals would be taken up at the expiration of the grant. UNH has almost become what I call PINO—Public In Name Only.

UNH also permanently ended many federal-grant-funded programs that protected the public health and maintained its own pipeline of science-savvy students by offering support to K–12 students and teachers. These included Watershed Watch (a program that trained students to monitor the health of watersheds), a professional development program for K–12 teachers in earth sciences, a STEM teachers academy, a summer camp that taught science to high schoolers, a STEM-teacher scholarship program, a program meant to develop American-citizen STEM professors, a program that included a study of how ecosystem scientists could better communicate their findings to policymakers, a program to bolster K–8 math teachers, classroom support for high-school chemistry teachers, the "Computer Science for All" initiative targeting K–12 students, a computing challenge for middle schoolers, a program to strengthen K–12 geoscience by developing curriculum and teachers, and a training program that allowed high-school science and math teachers to access better professional training and also borrow instruments to teach molecular biology, spectroscopy, molecular modeling, and chromatography.

Perhaps predictably, the lack of faith in New Hampshire's higher-educational institutions trickled down to its K–12 public school system. Between 2014 and 2024, roughly 18,500 students, or 10 percent of the total, opted out, many choosing home schooling instead. In some public districts, students who remain on the official rolls are barely present—in 2023, one district reported that nearly a third of its students were "chronically absent."

While New Hampshire is an outlier, it is not alone. It might soon be considered America's first de-institutionalized state, but it won't be the last.

Nationwide, over the last twenty or so years, America lost 136 rural hospitals, 861 colleges, 9,499 higher-education campuses, and tens of thousands of churches. In 2018, one of the last American institutions to hold high levels of trust from the American people was Big Tech; by 2021, Google, Facebook, and Amazon had lost more trust than any other institution over the same time period, and ranked lower than local governments or police.

What's an institution to do?

Laundry lists of experts have offered laundry lists of ways that America can rekindle its belief in itself—some say it is re-creating a culture of American pride; others focus on strategies like leveraging the remaining pockets of trust people feel for neighbors and businesses or building a more well-informed populace through boosting local newspapers and emphasizing civic values in schools. Still others call for political reforms like getting money out of politics, addressing income inequality, increasing transparency, and adopting new models of policymaking that empower the voice of those affected by those policies. All of these long-standing ideas have one thing in common: they either ignore paranormal beliefs altogether, or label them as conspiracy theories that cannot be given weight or consideration by decision-makers.

I would invite those decision-makers and analysts to try a different strategy: consider the jinn.

Belief in these spiritual entities, and their ability to possess humans, is widespread in certain Muslim populations. A majority of college students in Saudi Arabia believe jinn are the root cause of epilepsy, and a different study found that more than two in five Muslim psychiatric patients in the Netherlands thought that their aggression, depression, and hallucinations were caused by jinn.

For this reason, many faithful Muslims will whisper a quick prayer as they enter a bathroom, to protect themselves against the possibility of urinating on an invisible jinn and thereby earning its disfavor.

Eliz Hale, a social-justice activist and religious-studies PhD candidate at the University of California at Santa Barbara, has researched jinn possession, which results in symptoms that Western medicine classifies as pathological behavior.

Hale looked at different modes of treatment for jinn possession. She found that, for many people suffering clinically diagnosed PTSD, depression, and schizophrenia, a jinn exorcism was a very effective treatment. During the ceremony, healers known as *raqi* engage in dialogue with the jinn and expel it by using language from the Quran; the words of a passage are sometimes written on paper that is dissolved in water and then drunk by the possessed (and hopefully, when pissed out, does not provoke the next jinn).

In other cases, traditional treatments for jinn possession fail completely. Hale said that this is often because there is an obvious physical cause, such as infection or softened brain tissue, which responds only to evidence-based treatments.

The best approach is obvious, according to Hale.

"I would suggest adopting a philosophy of pragmatism," she said. This means providing an institutional treatment in which the underlying spiritual belief is not discredited, and including *raqi* exorcisms as a viable treatment option. And this is exactly what many medical clinics in the Muslim world offer.

But Hale's paper was not simply about jinn. She compared and contrasted jinn possessions with alien abductions.

Aliens, like jinn (and spirits, for that matter), are thought of as powerful and intelligent creatures that exercise free will and can be helpful or harmful to humans. And in both cases, the paranormal experience can leave a person with symptoms of mental disorders.

The main difference, Hale found, is that alien abductees don't have the treatment options that are available to those who are possessed by jinn.

"Mental health professionals who discount such beliefs as mere delusions are dismissing experiences which, for some people, constitute part of their lived beliefs and realities," Hale writes. "To do so is not only disrespectful to their patients, but damaging to their own ability to help provide effective care for people who have had possession or abduction experience."

The historical and religious context of the jinn allows the medical establishment to see it as a matter of cultural sensitivity; the lack of this context allows the establishment to view abductees as simply delusional.

Mike Stevens sees the impact of this during his work with abductees.

"A lot of them would rather be crazy," Mike said. "Because we have medicine for crazy. We don't have anything for this."

Hale argues that alien abductees should be treated with three critical components employed by *raqi*—curry the patient's faith in the treatment, provide group peer support, and allow for a persistent connection with the supernatural entity moving forward.

She specifically recommends group peer support and hypnosis, exactly the sorts of treatment that have evolved under Mike and others in New Hampshire's experiencer community.

Joel Blackstock is a social worker in Alabama who, dissatisfied with the state of America's medical systems, became the clinical director of a complex trauma collective in Birmingham. He's adopted a strategy of dealing with those who identify their trauma with various conspiracy theories or alien abduction. He said Hale's theories bear out in practice.

"A lot of the time when you're dealing with people who have a belief that's just not true . . . you work with it like you would work with a dream or a psychotic episode," Blackstock said. "The stuff that they're telling you is true to them. It's a metaphor for their emotion or a myth. It is emotionally true. You have to take it at face value and figure out what it means. If you're able to hold that then the people will have insight a lot of the time later that this is just how they process pain. Getting into a debate or fighting with them about what's real is a waste of time. You need to follow the emotion and help them heal."

It's the question of individual experience versus evidence-based science all over again. As with hypnosis and table tipping, the belief of efficacy creates efficacy.

This approach doesn't just apply to how medical science ought to treat jinn, or aliens. Barring a momentous Solid Phantom discovery by a group like KRI, it's clear that public beliefs in the paranormal will always far outstrip the evidence.

With that gap poised to widen, institutions need to undergo a fundamental change. In addition to earning back the public's trust through a renewed dedication to the public interest, institutions should stop pretending that our new American spiritualism can be ignored, derided, or debunked out of existence.

They need to instead follow the blueprint laid out by jinn possession treatments—to grapple with paranormal beliefs at a point of intersection between those beliefs and evidence. This means establishing a base level of mutual respect every time spiritualists come into contact with institutional apparatuses. Supporters of evidence-based policy might worry that creating paranormal-friendly spaces will strengthen false beliefs—but those supporters should consider that the evidence suggests an institutional stamp of approval might weaken those beliefs instead.

Institutions that want to influence believers should consider the teachings of academics like Melissa F. Lavin, associate professor of sociology at SUNY Oneonta, who studies populations of people that are stigmatized as deviants.

Lavin says that mediums, psychics, and tarot card readers are a lot like sex workers. Both groups are comprised of independent contractors, working outside of mainstream constructs, in activities that are both marginalized and stigmatized. And, as with sex workers, the customer is less stigmatized than the practitioner.

One impact of stigmatization is that practitioners are left to forge their own identities—as life coaches, as healers, as counselors and therapists. This positioning allows them to provide services in competition with institutions; Lavin said areas that are underserved by institutional medical care unwittingly provide fertile ground for mediums and psychics offering healing services. And the widespread provision of these alternative services can mask the need for institutional services in certain areas.

Lavin has traced destigmatization campaigns of various "deviant" groups—marijuana users, queer marriage partners, people with neuro-divergences like ADHD, and people with high levels of stigmatized features, like shortness, fatness, or ugliness.

In the not-so-distant past, a short, fat, ugly man with ADHD, a pot habit, and a husband would have been hampered by the societal obstacles in his way, which would have included social ostracism and legal peril at the very least.

But today, all those attributes have been at least somewhat destigmatized by expanding what is normal. Sometimes, normalization comes in the form of movies, music, and television that embrace those traits; sometimes it is legal protections; and sometimes it comes though recasting a behavior or condition as a medical feature rather than a flaw.

Destigmatizing mediums, tarot readers, psychics, and astrologers would reinforce their role in society, but it would also give public institutions an opportunity to define parameters for their practices.

I asked Lavin if certifying psychics, for example, might allow someone like Beau to practice mediumship ethically, and help isolate destructive tarot readers like Seth Mazzaglia. She responded that, in our "credential society," many intuitives would see certification as appealing insofar as it helps them to gain mainstream respect and legitimacy.

"Normalization of Tarot labor and consumption will continue to include a market for certification and other attempts at gatekeeping," she wrote. "For better or for worse."

Traditionally, government institutions have failed to recognize things like mediumship, in part because they didn't want to empower a worldview that lacked scientific evidence. But Lavin's research suggests that one way to defang the threatening aspects of mediumship would be to certify it as a recognized practice.

Thirty-One

HUMOR AND FREQUENT LOVE-MAKING!

Oh Abby, with all the mistakes I made
you love me still
The instant I cried for you, you returned to me
as though no time had passed
I see now that grief itself was the error!
　　　　—Stephen Sakellarios, *Loving Abby in Truth and Spirit*

"Welcome, everybody who came out here tonight."

The white light of the projector pulled parts of Val from the shadows of the Center's big meeting room. Four years ago, when Antje and Beau left KRI, Val worried. Now, she was grateful they had gone. She understood why her intuition had told her to stick around. Antje and Beau were like two large stones, filling the Center with a psychic weight and personality that displaced Val to the watery margins. When they left, she was sucked into the Center's center, and given space to grow.

Now, as KRI's most senior intuitive, she led the Center's meditation sessions and handled its spiritual security. Her empath support group had attracted new students, including a man named Dave. She asked him out.

Andy and Mike called him "CIA Dave," because they, only semi-humorously, suspected the CIA had sent him to monitor them.

The Center as a whole was not quite as vibrant as it was when Beau and Antje were contributing, but the search for a Solid Phantom was surviving, in part because Andy had turned one of the downstairs offices into a massage and Reiki parlor. In its small waiting room, there was a jar of "Spiritual Stones," which invited people to take out a rock, and put in cash. Andy was also exploring the possibility of launching a local television show, *Exeter Terrestrial*. And they had continued their commitment to bringing in unique speakers with interesting ideas.

Val was introducing one such speaker. Stephen Sakellarios was tall, bearded, and balding, wearing a vertically striped buttoned shirt over belted khaki pants that described his personality in gentle murmurs. He had a master in counseling and human systems, and worked as a legal transcriptionist from his attic studio apartment in Portland, Maine. He presented as blandly inoffensive, but he was actually a bowl of oatmeal with a bear trap of provocative beliefs buried inside.

Sakellarios's passion was reincarnation.

More than one in four Americans believe in reincarnation; studies of past-life regressions created a rare point of agreement between psychomanteum expert Raymond Moody and skeptic scientist Carl Sagan, who each found the results to be compelling enough to merit further study (though Sagan believed such research would uncover other explanations).

Other researchers have found that past-life memories accessed during hypnosis tend to include details of life that were not reflective of the historical period supposedly being accessed; rather, they reflected flawed views of those historical periods as portrayed in popular culture.

Val told the packed house that Sakellarios sought to prove reincarnation by matching past-life memories with actual historical people.

"In 2009, he began researching proposed matches with the most vigorous methods possible, given that the project was unfunded," Val said, adding as an aside, "God, do we know about that!"

The audience chuckled. Val said that, roughly 150 years ago, Sakellarios had also lived in Portland during a past life.

Aided by images projected onto the wall, he began to talk about reincarnation.

Taken at face value, there is an implausible incestuousness to reports of reincarnation, and nearly everyone seems to somehow connect with Napoleon Bonaparte. Napoleon himself believed that he was a reincarnation of the Roman emperor Charlemagne. Actress Shirley MacLaine believes she was a Moorish peasant girl who had sex with Charlemagne. Beatles musician John Lennon believed he *was* Napoleon, while World War II general George Patton believed he served as a soldier under Napoleon. Patton may have fought against actor Sylvester Stallone, who believed he was a soldier who was guillotined by Napoleon's forces (Stallone also said he lived a life as, perhaps, a Guatemalan monkey). Lennon also thought he was Jesus, and Patton also thought he was the Roman soldier who pierced Jesus's heart with a spear.

For those keeping notes, this means a single soul would have spent time on Earth as John Lennon, Napoleon, Charlemagne, and Jesus. Clearly, this is a field that cries out for evidence. Because the evidence for reincarnation is invariably rooted in individual experiences like memories and visions, it's almost impossible to reconcile it with the standards for science-supported facts.

Sakellarios, who had been studying past lives since 1973, said that it's easy to prove your reincarnation to yourself, but difficult to prove it to others. However, Sakellarios says that he has found a way to bridge the two, at least for a small subset of reported cases that meet rigorous criteria. The key, he said, was to painstakingly document one's past-life memories, and then seek a match to an actual person. But in order to meet the burden of proof, your past self must occupy a narrow zone of historical prominence—documented, but not famous.

"You need a past life that's very obscure, but you can find it if you keep digging," said Sakellarios. "It has to be available, but not readily available. And mine happens to fit that."

Sakellarios told the audience that he had previously walked the earth as Mathew Franklin Whittier, an obscure Civil War–era New England satirist (and brother of the better-known writer John Greenleaf Whittier).

One of Sakellarios's key memories came from an 1800s-era sexual encounter with Whittier's wife, Abby, in the home of her parents. She led him partway up the stairs and then paused on a landing.

"Halfway up on the right is a door cut into the wall and it looks like a white ship's hatch," he said, describing the memory to the KRI audience. "And when we get up there, she opens that door and pulls me in. And there's a little room, like a big linen closet or a little room, in there, and we make love in there."

The memory is significant to him because he eventually found the house in which it had taken place. In Haverhill, Massachusetts, he discovered that Whittier and Abby's former home included the staircase he remembered.

He showed the KRI audience projected images. Near the staircase top there was a split, with three stairs going up to the left, and three to the right. The left led to the second floor, while the right led only to a large closet.

"I couldn't possibly have ever seen it before," he said. "I saw something that there's no possible way I could have known about."

Sakellarios's claims go far beyond having merely lived as Whittier. His unique window into Whittier's life has put him dramatically at odds with the official historical record on Whittier's literary accomplishments.

When Whittier died in 1883 at the age of seventy, he was best known for a modest body of work—roughly seventy stories about Ethan Spike, a bombastic character from the fictional town of Hornby, located in Oxford County, Maine. Spike represented the most extreme (not to say fictional) traits of the rural Yankee stereotype—a mix of twisted dialectical syntax, idiotic misunderstandings of simple words, patriotism, grit, and a can-do attitude.

In one typically comedic bit, Whittier's Spike convinces the governor of Maine to guard against the possibility of a Confederate South warship mounting an attack on landlocked Hornby.

"Wal, sposin she is—haow is she goin to get to Hornby?"

"By the canawl," says I.

"But haow can she get through the locks?"

"Isril," says I, "she'll pick 'em!"

"God bless my soul!" says he, "I never thought of that, I'll call a caounsel meetin tonight."

This sort of wit is why Whittier has not earned much attention from academics. But Sakellarios says that the true Whittier canon is far more extensive. He claims that Whittier and his wife, Abby, were activists supporting the Underground Railroad, a politically delicate position that drove Whittier to write under pseudonyms. Hundreds of pseudonyms, beginning when he was a boy of twenty.

"Over ten years," Sakellarios said, "I've found 2,400 of his published works in all genres. Poetry, essays, editorials, everything. Reviews."

If this is true, then Whittier significantly outproduced the combined efforts of his literary peers Elizabeth Barrett Browning, Charles Dickens, Margaret Fuller, and Edgar Allan Poe. And this is significant, because Sakellarios says that Whittier's work was plagiarized and pilfered by all four of those writers. He believes that his former self was the true author of, among other works, *A Christmas Carol* and *The Raven*.

"None of this," Sakellarios reminded his audience, "you'll find in the official Whittier history."

What I found really striking about Sakellarios is not his beliefs, but the extent to which he has tried to document them. In 2014, he self-published a book called *Mathew Franklin Whittier in His Own Words*. The book's Amazon Kindle listing warns that "Due to its large file size, this book may take longer to download." The listing says it is 3,850 pages long. To visualize how much information that is, go buy seventeen copies of this book, and stack them up on top of each other. Full-priced hardcovers, please.

And yet, that epically sized book doesn't cover all Sakellarios has to say on the subject. He wrote a follow-up missive at a comparatively slim 1,816 pages, and then eight more books, all covering different facets of Whittier's life. Sakellarios has also produced academic papers, magazine submissions, documentary pieces, professionally disseminated press releases, and blog posts. He's reached out to hundreds of literary scholars in an effort to convince people of Whittier's vast body of work. It all adds up to, if not the most well-documented case of reincarnation, at least the most-documented case of reincarnation.

And yet, his efforts to gain traction in academic circles proved fruitless.

"Nobody takes me seriously about any of this stuff," he said. "I'm waiting for the day somebody does because this is a very significant historical find."

He's uncovered other details about Whittier, also unsubstantiated by scholars. Whittier, he is sure, was heavily engaged in the occult, and was also interested in reincarnation. He says that Whittier encoded messages in his work, some of which were intended for his future self—Sakellarios. Other messages were, he believes, for posterity.

"Well, the only person in posterity that ever figured it out was myself," Sakellarios told the crowd.

This is a very strange bag of cats, but the meows get even more peculiar.

Soon after establishing his connection to Whittier and beginning to research his former life, Sakellarios sometimes wondered about Whittier's wife, Abby, with whom Sakellarios remembered various romantic and sexual experiences. He thought of her often. In whose body was her soul currently residing? What would she look like? Would she remember their former life together? Or would she consider him to be a stranger?

His interest in Abby took a turn in March 2010, when a medium conveyed a message to Sakellarios from her spirit, who resided in the astral realm. Though she was not on Earth, she wanted to reconnect with him. Abby's unincarnated status seemed to create an unbridgeable chasm between the two. At first Sakellarios wasn't even confident that she was really hovering invisibly in his periphery. But over time, a convincing pile of evidence accrued—at one point, he dreamed of her, which drove him to scour the internet and collect about a dozen images of portraits of similar-looking women. He eventually came across an actual period portrait painted by a relative of Abby, and the portrait seemed to match the other images he had collected. Another time, he noticed that his two cats were staring at a corner of his living room. Could they see Abby?

He spoke to her, and asked her to move to a different corner of the room, and in a moment, the cats transferred their attention accordingly.

That's real evidence, he thought.

But he still had doubts. Was Abby really there? Or was it just his own wishful thinking?

He soon struck upon a method of direct communication with Abby that he termed "a kind of divination." Her spirit had a particular "energy signature" that he thought he could sense.

"Once I recognized it," he said, "she could pulse it."

He and Abby began to play a game intermittently. Sakellarios would sit in his car, eyes closed, and let his fingers wander over his CD collection. When Abby pulsed, he chose the CD he had been touching. He would then consider each track by number, and let her guide him to a specific song. Embedded within that song would be a message, from her to him.

It was a way of reassuring himself that Abby was not just a figment of his imagination. "Often it was responsive, something that I was writing or that was bothering me, and she would answer," he said.

Finally, Abby chose just the right song, at just the right time, to put his doubts firmly to rest.

"It affected me so deeply that I knew it was real communication," he told the KRI crowd. "I had been writing about remembering us making love outdoors, on picnics. Okay?"

Sakellarios played the song that Abby had chosen in response to his sex writings. It was "Starlight," from the album *Stars/Time/Bubbles/Love*, released in 1970 by The Free Design. Sakellarios knew it well.

The lyrics mention the sky and the wind, and reference being "under starlight" and having "touched your sacred skin."

This was, to Sakellarios, deeply moving and compelling evidence.

"Star, starlight you ever send," the song concluded. "Love, if night could never end."

This is real, he said to himself. *This is responsive. And she loves me.*

Even as he related it to the KRI crowd, years later, he blew air from his lungs, as if expelling heavy emotions. "So I never questioned it again after that."

With this knowledge, Sakellarios was thrilled to fully commit to a long-term relationship with Abby's spirit. Though he would never enjoy a physical embrace with her, he had found someone more perfect than any flesh-encumbered partner could ever be. Sakellarios's identity had become so entwined with his beliefs about reincarnation that it would be laughable to think that he could be dissuaded from them by any method of debunking or persuasion in the institutional toolbox.

He acted as any love-besotted person might. He wrote her impassioned poems on holidays. They watched movies together. He disavowed having doubted her existence, and messing around with other women. In a book that tells the story of their reunion, he wrote glowingly of a love rekindled after 170 years.

"When we were physically together," he wrote, "we soared on wings of intellectual sharing, joint creative and philanthropic measures, humor and frequent love-making!" He placed an exclamation point after the phrase love-making, which I assume means it was really good.

At the end of his speech, he received a round of applause. KRI had delivered another win to their patrons, and would survive another day.

Who doesn't enjoy a good love story?

One day, not long after the Sakellarios presentation, the landlord did a walkthrough of the KRI offices with Andy.

"Oh my God," said the landlord, sniffing the air. "Is that the cat box?"

"No," said Andy. "Remember? I had to fix the carpeting? I told you I was going to take the glue up and save the carpeting. That's mold. It's mildew from under the carpets. Don't worry, I'm going to steam clean it."

But Andy was lying. The smell was coming from Mike's room.

Mike had attained a position of respect and love that few alien abductees could ever hope for. He had successfully fought for the state to erect the marker commemorating the Hills. He was a regular fixture at the well-attended annual Exeter UFO Festival, at which he gave guided tours of the locations significant to the Exeter Incident. He had rallied a community of experiencers for mutual comfort and support. He'd displayed kindness, quiet strength, and a willingness to help with tasks great and small. He was loved.

But he was not doing well.

Humor and Frequent Love-Making!

In the early summer, some of his coworkers at Troy's flooring company had made fun of his interest in aliens behind his back. The "jabbing and jarring," characterized as jokes, got to him, so he quit. Rather than seek a new job, he spent the summer focusing on his volunteer work within the UFO community.

Soon after, Andy and Val learned that he was sleeping in his van in the KRI parking lot. Andy rented Mike a basement-level utility closet that used to hold wiring, at a cost of fifty dollars a week. It was furnished with a dresser, which Mike began to fill with empty liquor bottles.

"I'm not kidding," said Andy. "He carpeted it with old pizza boxes. It was that revolting. And they weren't empty pizza boxes. Sometimes they were, sometimes they weren't. But they were great for soaking up the beer he spilled."

Things between Mike and Andy were getting a bit tense. For a while, Mike dated a woman that neither Val nor Andy liked. Val felt like the woman was actively working to chill her long-standing friendship with Mike, while Andy found her to be unattractive and unintelligent and annoying and a poor medium. For starters.

Once, Andy was in his office, talking shit about Mike's girlfriend to Beau's ex-husband, Troy. Andy opened the door to find that Mike was standing right outside, listening.

"How much of that did you hear, Mike?" asked Andy.

"Enough," said Mike.

"Sorry, dude," said Andy. "It's just an opinion."

Mike said nothing. Andy couldn't tell whether he was deeply offended, or took it in stride.

"When you've got a talkative person in front of a quiet one," Andy said, "nothing changes when he stops talking to you."

Exeter's UFO Festival, held at the end of August, was a big boost to Mike's spirits. One reason was that he got to hang out with friends like Michael and Michelle Mitchell. They were the husband and wife team behind Mitchell Comics, which had a body of New Hampshire–based paranormal-themed work. Michael Mitchell had created comics about the Exeter Incident and witches. He'd even included Mike as a minor character in *Granite State Bigfoot*.

Palling around with the Mitchells and the other alien-centric friends at the UFO Fest was an emotional lift for Mike, as it always was. But in its wake, Mike found himself feeling even lower. He hid his inner turmoil, and went through the motions. The Saturday after the festival, he helped organize a KRI class in spoon-bending, taught by an "energetic catalyst" from Bedford, Massachusetts. It cost ten bucks, plus a pair of socks to donate to charity. Beneath the surface, Mike also began to plan the details of the rest of his life. There was an old camping site that he knew of. He would bring pills, and beer. He would drink, take pills, drink some more. Then hang himself.

Thirty-Two

THE UNEXPECTED PASSING OF MICHAEL

> *I have an idea that Gatsby himself didn't believe it would come, and perhaps he no longer cared. If that was true he must have felt that he had lost the old warm world, paid a high price for living too long with a single dream. He must have looked up at an unfamiliar sky through frightening leaves and shivered as he found what a grotesque thing a rose is and how raw the sunlight was upon the scarcely created grass. A new world, material without being real, where poor ghosts, breathing dreams like air, drifted fortuitously about . . . like that ashen, fantastic figure gliding toward him through the amorphous trees.*
> —F. Scott Fitzgerald, *The Great Gatsby*

When the seemingly impossible news circulated, it left a sense of grief and loss throughout the broader KRI community. Death may have added a spirit to the hereafter, but it was the hole left in the land of the living that stung.

Antje was by then well established as a medium in the office she rented in Newmarket. Here, she freely crafted her own protections from liability, in both the legal and the spiritual sense. One day, she looked across the threshold of her doorway and saw a clear-as-day image of Mike Stevens, tall and quiet, looking at her with those soulful eyes.

This was no spirit. It was Mike, in flesh and blood and fully as alive as she was. He didn't tell her how he had survived his planned fireside suicide. He didn't even tell her that he had experienced suicidal thoughts at all. He'd bottomed out on Sunday evening, the day after the spoon-bending class. He was lying on his mattress in the KRI utility closet, scrolling through his phone and thinking grave thoughts. When he went to that final campsite, he thought, he would drink Elysian Space Dust, his favorite beer. That's when he noticed a Facebook post from Michelle Mitchell, of Mitchell Comics.

"It's with heavy hearts that we share with you the unexpected passing of Michael," she wrote. "He inspired many by sharing his art and love for creating. His legacy will carry on through his books, illustrations, stories, and memories shared by his friends and loved ones."

Over the next couple of hours, Mike watched in real time as the paranormal community responded to the tragic news.

"This is a truly devastating loss for our world," wrote Val, in one of literally hundreds of heartfelt posts. "Michael has such a genuinely kind soul. So glad he will live on in his comics as well as the hearts of all of us who knew him."

Scientists have shown that the loss of a loved one can trigger suicides, resulting in death rates between two to four times as often as among those who have not been bereaved. But as Mike, who had been actively planning his death, scrolled, he saw the outpouring of grief from the community. The experiencers—the very people Mike had devoted himself to protecting and uplifting—were devastated at the loss of Mitchell.

That held power for Mike.

"This is what you're going to do to other people if you go through with it," Mike told himself. Confronted with the pain of death, Mike chose to live.

Val and Andy were completely unaware of the life-and-death drama playing out in Mike's head. Val could tell he was going through a difficult time, but she had no idea it had gone that far. He hatched, nurtured, and buried his suicide plan, all without their knowledge.

When Val found out, she blamed Andy for not being a better friend to him. She blamed herself, too.

Andy had always been an advocate for Val at the Center, and when Beau left, their bond of friendship had become even stronger. She'd come into her own partly because he relied on her so completely to provide spiritual leadership to the community.

As part of that role, Val ran a regular KRI event called Lifting Lodge, during which attendees sat in silent meditation for up to forty-five minutes. As soon as they were ready, Val told them a message she had received from their spirit guides, about their path.

When Andy sat in on a Lifting Lodge, she brought him a message from beyond.

"Utilize the community that the Center has created to start doing your research," Val told him.

Andy didn't agree with the premise of the message. Start doing research? He had started years ago!

"He kept saying 'I am I am,'" said Val. "But he wasn't."

Val was getting impatient. Andy had talked, for years, about doing more formal scientific research of the paranormal. They had a handful of accomplishments—Andy's book, a set of superior ghost-hunting protocols, the orb experiments, and the raccoon video. But mostly, it amounted to a heap of personally compelling anecdotes. There was no school, no lab, no double-blind experiment, no submission to peer-reviewed journals. Andy's grand ambitions had taken a consistent back seat to renovations, and UNH coursework, and flooring work, and various other distractions. Where was the actual science?

"Something seemed to be holding him back," Val said.

She was also irked about the financial status of her two paranormal-themed books. Andy had edited and published them under the Runestone label, which he had acquired from Beau's medium friend. Val had no idea how many books had sold online, or how much they had made. And no matter how vocally she raised the subject, there were no answers to be had. Andy wasn't accounting the accounts.

"The money was going to Andy, but he couldn't figure out how to track it," she said. "I never saw a dime of the money other than what I sold myself in hard copies."

Val sensed that the strain of KRI's finances was beginning to wear on Andy, but he wasn't open about it. Until one day in October, about six weeks after the UFO Fest, when Andy decided that he'd had enough.

"We're done," he told Val and Mike. "We're closing."

A decade of barely keeping afloat had stressed him more than even he realized. As soon as he made the decision, a perpetual knot in his chest slipped its own bonds.

"The ending of it was the single most relieving thing of my life," he said.

Andy took responsibility for flying by the seat of his pants.

"I was always several months behind the level of preparedness I should have been at for any given situation," said Andy. "I went in, just wung it, and I was winging it the whole time."

He acted swiftly. Just weeks later, at the end of November, Andy moved into an apartment, and the Center closed its doors. Forever.

There would be no KRI school. There would be no Solid Phantom, at least not through the Center.

Val wanted to continue her empath support groups, which had become very meaningful to her. She tried holding her first at a Starbucks in Epping. Val held a chai in one hand and a croissant in the other while six or seven empaths crowded around a big table, holding steaming cups and sharing stories of feeling the feelings of others.

But Starbucks was crowded, and Val couldn't control the emotional energy of the space. It was difficult to find the stillness that facilitates communication with an inner voice, when people in the same room were ordering grande lattes. Val knew she needed somewhere else.

She floated through a couple of other locations—a dedicated workspace in the home of a spiritual energy worker, and then a small crystal shop in Stratham. But neither of those worked out. She kept looking.

For Mike, KRI was not only the central location of his UFO-related activities. Its closure meant he would be unhoused, or at least un-utility-closeted.

The Unexpected Passing of Michael

One night, when Andy collected the cash box, he found that it was missing a hundred-dollar check that he had seen there the previous night. Andy said that, when he was on the brink of calling the police to report the missing money, Mike admitted that he had taken the check.

Andy said he was floored.

"What are you talking about?" he said, describing it to me later. "I never imagined our relationship had degraded that way."

Mike was not confrontational.

"He said as little as possible," said Andy. "He offered to give me twenty dollars of it back."

"He was on a downward spiral but hadn't hit rock bottom yet," said Val. "I think stealing from Andy was rock bottom for him."

These were the circumstances that led Mike to knock on Antje's door. He didn't ask for a place to sleep. But he needed a location for his abductee support work. To him, that had become everything.

Could Antje help?

She threw him a set of keys, and told him to move his entire KRI desk and work items into an unused portion of the rear of the unit. "I would do anything for him," she later told me. "He's a lovely person. He's great." She told Mike he could use a rear room as an office, to hold group meetings—whatever. With space for his abductee work secure, Mike wound up living in a combination of the woods and his minivan. When he wanted to stretch out, he opened a window.

A minivan is not designed to be slept in, but if you're in a pinch, you can make it work. And science is not designed to accommodate views on Bigfoot, aliens, or spirit guides. Or perhaps it's the paranormal that's not designed to work with science.

It's an inherently uncomfortable juxtaposition. Paranormalism is a big, wild man crammed into the scientifically engineered confines of a minivan.

Though, in a pinch, you can make that work too.

Thirty-Three

ALMOST CERTAINLY COMPLETELY FALSE

> *"It's dead, whatever it is," said Holmes. "We've laid the family ghost once and forever."*
> —Arthur Conan Doyle, *The Hound of the Baskervilles*

When Beau and Andy first met, they were at the forefront of a movement of New Age spiritualism that is reshaping America. Millions of people are rejecting institutional knowledge and authority in favor of a reality defined by ghosts, mediumship, aliens, and cryptids.

Individuals within the movement are fueled by formative personal experiences that they find meaningful—Andy's dead father's voice on a recorder, Beau's contact with a menacing alien in a bar, Mike's childhood abduction, Val's ghostly sexual assault, and Antje's Tupperware containers eerily lined up on her kitchen floor. All these became lifelong touchstones that shaped their identities.

Institutions have, in their institutional way, been extremely slow to understand a shift that is rooted outside of evidence-based knowledge. And they've been even slower to react.

To survive, institutions need to mount two simultaneous, and fundamental, reforms. First, they must earn back the public trust with bold and transparent initiatives that serve the public good, whether that means providing a meaningful education, reducing income inequality, holding elites accountable, administering justice fairly, or inviting the public more deeply into the scientific establishment.

And secondly, institutions need to maintain the sanctity of evidence-based knowledge while appealing to the many citizens who place a high value on other ways of interpreting the world around us. This means treating paranormal beliefs with the same delicate level of respect that has traditionally been extended to major world religions—culturally inclusive spaces in medical settings, grant-supported community centers that can direct the energy behind spiritualism toward the public good, and grassroots-informed certification programs for mediums, ghost hunters, and other paranormal professionals, among others.

These reforms seem unlikely, especially in the wake of the 2024 reelection of Donald Trump, who embodied a sense of institutional destruction and instability. Research professor Christopher Bader's annual Chapman University Survey of American Fears found that paranormal beliefs spiked in the years following his first presidential term. By 2019, a majority of Americans, roughly 55 percent, believed in haunted places, and roughly 65 percent believed in Atlantis, up from a minority of Americans just a few years earlier. Bader attributed this to a shift from institutional to personal belief structures.

It is likely that, rather than fundamental reform, America will continue down the path that has defined the past thirty-odd years: dysfunction, cultural isolation of institutional power structures, corporate-led democracy, and an inability to adapt to a shifting world paradigm.

This means New Age spiritualists will continue to advance into the void left by shrinking institutions. And how that plays out will rest in the hands of people like Beau and Andy, who have retreated into different corners of the spiritualist belief systems.

During the 2023 fall semester at UNH, I parked on campus, and threaded my way through the buildings and scores of young students who maintain faith that college institutions have something to offer. The state of New Hampshire continued to be a national leader in the business of distrusting one's neighbors. Among its many distinctions related to that category was a *Forbes* magazine study that in 2023 ranked the Granite State the first in the nation for package-nabbing porch pirates, perhaps a nod to the state's piratical seafaring history.

I headed into McConnell Hall and walked past the office of Cliff Brown, who has grappled with the loss of graduate students and other departmental resources. I threaded through long corridors, past vending machines and bathrooms and water fountains. Finally, I walked into a class full of students busily scribbling the answers to a pre-break exam, and sat down at the front of the room, next to their teacher.

It was Andy Kitt, whose status as a PhD candidate made him a valuable asset to the psychology department. Andy is close to joining the ranks of those whose work has shed light on hypnosis and the Mayan calendar, on evolution and climate change, on the speech of birds and whales, on the bizarre properties of lightning and the moon and black holes and oh-so-many tangible things: those UNH academics include hypnotism expert Ronald Shor; archaeologist William Saturno; biologist Randy Olson; climate change impact experts Elizabeth Burakowski and Matthew Magnusson; UNH avian communication specialist Karina Sanchez; UNH marine acoustic ecologist Michelle Fournet; lunar astronomer Andrew Jordan; gamma-ray-astronomy professor Joseph Dwyer; and UNH astronomer Dacheng Lin.

And soon, psychophysics expert Dr. Andy Kitt. For now, Andy was teaching two sections of a statistics class. On off weeks, he did flooring work for Beau's ex-husband.

"It's not my job to teach you how to use a calculator. It's my job to teach you the mental framework, the underlying framework," Andy said. He had finally given up smoking with the help of a Russian-born long-practicing hypnotist whose approach to combating various sorts of addiction included bonking the addict on the head.

When students approached him with hushed questions about a question, he questioned them back, steering them toward the correct answer. It almost always worked.

"So how do you find out what percent isn't between those two numbers? Did you find the percent from the p?"

Behind the lectern, Andy shone with magnanimity. On instructor feedback forms, his students cited his passion, caring, geekiness, and propensity for baking cookies on the last day of class.

"I try to make tests so the first person done will finish about halfway through the class period. It makes for a fair test," he told me. "It means people who got it easy get out fast. The people who have it hard have twice as much time to do it."

But this is still, recognizably, Andy.

After Andy and I left campus to get lunch at a Seacoast restaurant, he practically leapt from his chair, goggling out the window at someone walking on the sidewalk.

"I told you about the really short dude, the guy from KRI who was a dwarf that we used to make fun of because he was short, who was cool about it?" He hadn't told me. "That's him right there."

Andy craned his neck to get a better look at the man who was shorter than he, before he disappeared from view.

"I feel bad. I wasn't making fun of him," he chuckled. "I was just making a point, you know? 'Cause I had—there's another guy at KRI who's really short."

"Uh-huh," I said.

"And I was making jokes. It's like: 'Dude, you're not even close to the shortest guy. I've had a guy in here who was, who, who literally suffered from dwarfism.' So he's the shortest guy ever that I've spoken to."

Andy motioned to the window, still tickled.

"There he is," he told me. "That was him walking by. That's so funny. I can't remember his freaking name. It kills me."

For years, Andy has felt that an earned PhD would allow him to conduct psychophysics experiments that would get paranormal knowledge past the gatekeepers of academic scientific institutions. He planned to use that legitimacy to help him start a new Center, one that would more closely hew to his original vision of applying strict scientific standards to paranormal concepts.

Andy represents a wing of America's modern spiritualist movement that highly values the scientific process, and that seeks to correct flaws and biases within the scientific establishment by bringing evidence of the paranormal into existing scientific institutions.

This would put him in a small but growing group of serious academics who are in a position to unify the scientific and spiritualist worldviews. The instant a Solid Phantom is documented, ghosts will no longer be paranormal—just normal. If Andy's scientific approach, the one that is at least imitated by popular ghost-hunting television shows, prevails, it tethers the spiritualist movement to an agreed-upon set of facts.

Andy is positioned as a fair broker in part because of his willingness to reject individual experiences that lack evidence. And this came home to me in a very real way when, over the course of interviewing him for this book, he told me that he didn't believe in many of the seminal ghostly experiences that inspired the members of KRI. Antje's Tupperware experience? He rejected it based on a general assessment of her character. The Holos University study of Beau's powers? "The person conducting the study knew the outcomes at the midpoint, so it wasn't double blind." The menacing alien at the Coat of Arms pub? "Logically, it didn't work. She says this alien is going to bother us for talking about things we're already talking about." Mike's abduction as a small child? "The idea that he was abducted is almost certainly completely false." And the sexual assault at the Button Factory? "I don't want to badmouth any of the mediums or belittle their experience. They all had experiences." But, he said, they talked to one another, which robbed it of its evidentiary value.

And what of Andy himself, who entered ghost hunting after hearing his father's voice on Andy's recorder?

"Right," Andy said. "It was on a recorder, got transferred to a laptop, and then disappeared. And I've never been able to find that again."

If Andy represents the wing of the spiritualist movement that will preserve the role of science, Beau has become a leader in the camp that finds science to have only passing relevance to our lives. Part of her mission is to destigmatize spirituality.

"My whole thing is, no guru, no woo woo," said Beau. "I'm kind of demystifying the whole concept of intuition and trying to make it more normal."

When I spoke to her, she told me that, shortly after leaving KRI and moving to the Pacific Northwest, her computer business, the Idea Garage, "kind of fell apart." But her mediumship, and in particular her seven-week SAGE program, was providing her with a steady income.

By 2019, Beau had taught the SAGE course so many times that she knew exactly what questions students would ask, and when. This level of mastery allowed her to retool the curriculum so that students rarely had any questions that weren't addressed.

"In 2019, this yahoo here," she said, looking at a blank space in the air that held her spirit guide, "Peter, showed up and said: 'Put it online.'"

She quickly built a library of videos that would allow her to teach the course in pre-recorded sessions. When the Covid pandemic hit in early 2020, Beau had already digitized her offerings, and was charging $1,500 per student. Peter told her to reduce the price to $150.

"People were home and they were re-evaluating their lives, and it fit," said Beau. She was flooded with students from all over the globe. More than five thousand people signed up in the first four months (which, if you do the math, is $750,000).

"And then I had to kind of take a breather," Beau said.

She estimates that about a fifth of her clients join the "Sage Circle community," an online outgrowth of the program at a cost of $19 per month. "We do meditations and full moon stuff and all that kind of stuff together," Beau said.

She said that the number of new students has doubled, and continued to grow. Her clients are 99 percent women. Most are at least in their forties. Her oldest student, Rose, is ninety-two.

To meet the demand, she has assembled a network of staff members, all women, in New Hampshire, Vancouver, Oregon, Australia, and the Philippines.

It occurred to me that absent a Solid Phantom, one likely outcome from America's continued drift into spiritualism could be new loci of cultural power that fuse into something resembling our current way of life. As mediums, ghost hunters, empaths, cryptologists, and UFOlogists proliferate, they might organize into a new raft of organizations that serve the public, in much the way that chiropractors, acupuncturists, and yoga practitioners have slotted into the existing evidence-based health-care paradigm.

Beau's SAGE community is an early example of that: a post-institutional institution.

Currently, there are only two checks on these sorts of groups: they must operate within a legal framework, and they must appeal to market forces. That puts an extraordinary responsibility on the leaders of these groups to work on behalf of the public good.

Beau's SAGE Method embodies the best, and the worst, of what a post-institutional world might look like. The course has a heavy emphasis on troubleshooting blocks to intuition, which often involves the interior emotional baggage of the student. Its goal is to help people declutter their inner lives, connect with spirit guides, learn to focus, and manifest real-world benefits.

It also includes a certification course that holds instructors to a high ethical standard.

But the most striking thing about the SAGE Method is its avoidance of dogma, and its adaptability. It encourages students to experiment, and to find their own methods, rather than tie them to a specific rigid set of rituals; this is a model of tolerance, but also an abandonment of the notion of universal truths.

It supposes that, if tolerance and respect and inclusiveness are universal morals, then perhaps the need for universal facts is lessened.

When Beau used to say, back at KRI, that students should "only keep what resonates with you," it would set Andy's teeth on edge.

"That phrase is the antithesis of good science," he said. "She meant whatever you think is right, is right."

Just thinking about it made Andy say no. Many times.

"And it's like, no, no, no, no," he told me. "I don't, no, no, no, no. That's not how science works. The one that works is the one that works when you do it."

And that sums up the differences within the spiritualist movement that will have to be resolved.

There is a third path between these two worldviews, one in which science is neither preserved, nor rejected.

Michele Hanks, the New York University anthropologist who does ethnographic work with paranormal investigation groups, has found that the majority of ghost-hunting groups are not searching for evidence.

They use equipment that they don't know how to use, and cite physics principles that they don't understand. They're not citizen scientists, she said. They're actors, though actors who are thoroughly convinced by their own performances.

"The equipment helps them perform a convincing narrative that they are experts," she said.

She said that the universal trait shared by all popular pieces of ghost-hunting equipment—EMF detectors, white noise, indistinct recordings, thermometers—is their ambiguity. All generate data that invites speculation without answer.

And that's the point, said Hanks.

"The structure of amateur ghost hunting is designed to tease and maintain doubt, rather than pulling the individual toward a concrete camp of belief or disbelief. The unknown must be preserved," she wrote.

For the spiritualist movement to survive, it must embody both Sisyphus and Tantalus; the Solid Phantom must be just beyond the fingertips, enticing and titillating, only to withdraw every time a person reaches for it. The stone of evidence must be rolled toward a peak of enlightenment, only to slip from their grasp, again and again.

This is the perpetual journey that allows people to bypass brick-and-mortar institutions in favor of a utopian mirage; these are the steps that are changing America.

In this unending search for evidence, everything is something, but at the same time, nothing is anything.

Epilogue

And thou, a ghost, amid the entombing trees
Didst glide away. Only thine eyes remained.
They would not go—they never yet have gone.
       —Edgar Allan Poe, "To Helen"

Over seventy years, the University of New Hampshire Museum of Art amassed a collection of more than a thousand pieces, including paintings by Boston Expressionists Hyman Bloom and Karl Zerbe, photographs by Andy Warhol, etchings by Goya and Rembrandt, an extensive collection of ceremonial objects made in the 1930s by the Lozi people of Barotseland in south-central Africa, sculptures by Wendell Castle, and a bounty of two hundred "ukiyo-e," Japanese woodblock prints, some dating back to the early 1600s.

In 2021, the museum's director, Kristina L. Durocher, was elected president of the Association of Academic Museums and Galleries, the leading professional organization for academic museums, galleries, and collections in the United States. Under her leadership, the AAMG issued a statement in which it condemned the monetization of such art collections. To introduce a profit motive into art, it suggested, was a violation of the entire idea of bringing beauty and enlightenment to the world, including support for those who might not be in a position to pay for access.

The Ghost Lab

The high-minded ideals of the UNH museum were often put to the test. During the 2020 Covid pandemic, the museum responded by digitizing its collection and shifting to various sorts of digital interactions. During a budget crisis caused by a lower-than-expected 2022 enrollment, it beefed up its fundraising efforts, and raised roughly 15 percent of its $600,000 budget from donors and grants. The president of the board of directors, Linda Chestney, noted that the museum was well positioned to overcome a challenging environment.

"Adapting is a fine art," she wrote to the arts community. "We've mastered it."

Two years later, the museum closed. In 2024, a $14 million budget shortfall caused by yet another enrollment dip forced UNH to cut seventy-five more positions, and to close the art museum. Faculty said that, because UNH continues to receive the lowest state support in the nation, the shortfall of student tuition dollars had a disproportionate impact on the budget.

The closure of the UNH museum left Durocher in the uncomfortable position of being the continuing president of the national association, despite having had her own museum yanked out from beneath her.

"It is unusual for an R1 university and a flagship land grant university to be without a museum and yet," Durocher wrote drolly, "here we are."

The paintings, sculptures, woodblocks, ceremonial objects, and prints were placed into storage pending, I suppose, an age of rejuvenated public trust.

The last time I saw Antje, she was sitting in her Seacoast office (the paranormal one, not the paralegal one). We sat in comfortable chairs facing one another.

As a certified SAGE practitioner, she no longer gallivants around haunted houses in the dark. She does energy work, performs readings, and leads a Sacred Space event.

"Not tooting my own horn," said Antje, "but I have become very successful."

Epilogue

She has continued to have interesting spiritual encounters as she carries on the work that Beau trained her for. In addition to dictating that lost pickle recipe from beyond the grave, she has helped brewers locate the perfect yeast. She also clears homes on the market of negative energy for real estate agents. Business is good, she said, both in volume and morality-wise.

"I know this is a jerky thing for me to say, but I've done well almost in spite of Andy," she said. "I have to be extremely grateful for Andy. . . . I learned a lot about my own self-worth, not because he made me feel worthy, but because he made me feel the opposite."

The last time I saw Beau was via Zoom. She was similarly grateful for her time at the Center. The highs were incredible and the lows were instructive.

"If you're doing this book on the rise and the fall of the Center," said Beau, "it had potential. It fell apart. . . . I was not brave enough to put my foot down. I was his product." She took partial blame. "It started because of my shortcomings. It ended because of Andy's shortcomings."

Beau, who has operated mostly online for the last several years, is talking about opening a brick-and-mortar school dedicated to SAGE. With advice from her spirit guides, Beau is betting that the formula for future spirituality will be an institution without dogma.

It is the individual who matters.

"One path," she said, "isn't the only path."

Mike Stevens's profile continued to rise. He was the subject of a book written by a New England paranormal enthusiast, and he maintains a busy schedule of festivals and conventions, podcast appearances and support meetings for experiencers. Outside of KRI, he formed his own group, Granite Sky, to help experiencers through post-encounter trauma with his motto of "It's about people, not proof."

When I reached out to Mike, he was respectful, but declined to have even a preliminary off-the-record conversation about being interviewed for this book.

I finally came face-to-face with him for the first time at the 2023 Exeter UFO Festival, where I'd purchased tickets to a trolley tour that featured him talking about the Exeter Incident. When he was in between trolley rides, I introduced myself, hoping that seeing me in person might convince him to speak with me, but it didn't change anything. We made a bit of uncomfortable small talk.

During that same weekend festival, he was also a featured speaker for a lecture series in an auditorium at the town hall. He talked to an attentive crowd about media exploitation of UFO experiences. From his perspective, every sensationalistic television show or movie or book that tells the story of alien abductees is turning the whole phenomenon into a freak show, a for-profit plundering of a disadvantaged few.

This explained why he had rebuffed my advances.

"People have the balls to go out and purposely exploit," Mike told the crowd. "It needs to change. It needs to change. . . . Look at what we're doing. When it's done for the wrong reasons, by the wrong people, we're killing experiencers."

When Mike took questions from the audience, I raised my hand. I asked him what distinguished exploitative storytelling from acceptable storytelling.

"The big media are gonna be the ones that are just like, 'Give us all your stories,'" said Mike. ". . . Screw you. We don't care whether you like it. . . . We're not even going to let you tell us how we tell your story."

I wanted to tell Mike that giving a subject editorial control over a story wouldn't be journalism—How could we trust a media outlet where the news was controlled by the interviewees? But my response would have been the journalistic institution's answer to seeking objectivity by balancing truth and ethics. Mike's got a fundamentally different perspective, where the individual voice, no matter how subjective, reigns supreme.

Media is just one more institution that runs afoul of this idea.

In a post-institutional world, therapeutic support groups of peers, structured like an AA meeting, would be the best way to treat various

Epilogue

paranormal-associated illnesses, from demonic voices to jinn possession to abduction trauma.

Until that happens, Mike described the unique challenges faced by UFO abductees. Those abducted or violated by a person get access to evidence-based support services, and a search for justice by the legal system. Those abducted by aliens get, instead, to be made the butt of jokes. If they seek help in a therapeutic environment, they run the risk of having their experience labeled a delusion.

"This is a hard spot to be in no matter what got you there," said Mike. "Which again leads to shame and self-doubt. At that point, your chances of recovery on your own are near zilch."

Val also spoke at the 2023 Exeter UFO Festival.

"It was time to really tell my story," she said. A woman who had once quailed at the thought of being in a group setting mounted the stairs of a stage to address the large crowd and reveal, for the first time, that she, too, had been abducted by aliens as a child.

When she was four years old, she and her father stopped their car on the roadside and boarded a spacecraft that was inviting them in. Her father was frightened, but she found only a sense of joy and wonder in the radically inhumanoid inhabitants of the craft.

"Whatever we're inside of, it's a living entity, but not the same way that we're alive. It is more ethereal, more space between the parts that make it up and it has consciousness," she said. "Each piece has a consciousness."

Val has continued to be a leader in the spiritual community through her authorship, her public appearances, and her continued devotion to running a group for empaths, though in late 2023 she stopped characterizing it as a support group in favor of an empowerment group.

Val worked for a company that evaluated and rated different seafood harvesting operations, based on their environmental sustainability. She said the science she relies on in that capacity is fundamentally different from the science that underpins her ghost hunting.

"We have years and years, decades of scientific research for animals and the science behind growing them. That stuff is utilized in best practices for seafood," she said. "To try to establish best practices for paranormal practices, we need scientists like Andy. I want him to finish his degree already and see him do the work he talks about doing. We need parapsychologists. We need scientists dedicated to trying to find ways to quantify what is happening. I think a lot of people will agree something is happening. We think we know what it is. We need someone to figure it out."

Val did get a chuckle one day, when a woman came up after an empath meeting and said she had once gone to an event at the KRI Center.

"Who was that guy?" the woman asked. "Andy something."

Val replied that the woman was probably talking about Andy Kitt.

"I remember him introducing himself to me," said the woman. "He said, 'Hi. I'm Andy. I'm an asshole.'"

"That's him," said Val. "In a nutshell."

More ghosts crowd into the Seacoast region than ever before.

Where is Lizzi Marriott, whose body was dumped into the Piscataqua River? Where are Joseph Hesser, founder of Hesser College, and Dolly Markey, the humble alumna who outlived her alma mater of McIntosh College? Where is Peter Kitt? Where is Arthur Conan Doyle? Where are Ed and Lorraine Warren? Where are Betty and Barney Hill?

Their fleshly forms are in various states of decompose. Something is gone. Whether it has all gone into the bellies of a million tiny organisms that keep the world turning, or whether some portion has escaped into an ephemeral plane of existence that mortals can just barely tickle the belly of, is a matter of opinion. Or, if you prefer, fact.

Want to hunt ghosts?

Beau has some advice. "It would be really cool that if, when this book comes out, and anybody that was interested in creating a community,

Epilogue

took away from it that y'all have to grow," said Beau. "And everyone needs to, you know, find the better version of themselves if they're trying to help others find a better version of themselves. To keep moving forward in your hunt for ghosts, it absolutely is a critical ingredient."

Want to hunt ghosts?

Val has some advice, too.

"Go out there and do it. Do it well," she said. "Know the equipment. Validate things. Validate them well. Do it a few more times. Keep looking for better methods. Do your best. And have good intentions."

Want to hunt ghosts?

Andy Kitt has devised an experiment for you. It can be done by anyone in the world, with a single, inexpensive piece of equipment: a voice recorder.

If you want to do this, you can. Simply take the recorder to a quiet place in your home or office, garage or backyard, and open your mind to the possibility. Turn the recorder on. For ten minutes, ask questions to the air.

Do this for ten days running.

You are creating a small forum, one that gives voice to the most vulnerable, the most insubstantial population of them all—the deceased. The spirits will be drawn to your routine like flies to honey.

"Ask things of real interest," Andy urged me.

"What's an example of something of real interest?" I asked.

"Why are you here? Is there something that has drawn you to me? Just ask questions that, if you were dead, what would you want to answer? Like, what sorts of things did you enjoy in life?"

When you are done, go to another quiet place. Listen to your recording. If possible, pull up a video representation of the audio. Lose yourself in those little peaks and valleys that describe the sound of your recording. Each of your questions is a mountain range. But in the background hiss of the intermediate silences, you may detect a blink-and-you'll-miss-it anomaly. A modest hump among the mountaintops, a Hobbiton Hill among Andes. You will listen to it—once, twice, a dozen times, a trill sounding against your spine. You will write down a word or short phrase, then listen again. Then write another short phrase. You will ask others to listen, and tell you what they hear.

"I have never known anybody to do it more than ten days in a row without getting something on the recording that they couldn't explain," said Andy. "In each case, the responses I got were unambiguous."

Sometimes, Andy says, you will hear from a specific person that you knew in life. Other times you will hear things that make no sense, sounds generated by an unknown being in an unknown place at an unknown time for an unknown reason.

That tiny sound half-buried beneath a delicate hiss is the noise of a mind cracking open. A mind that is ready to reject the scientific paradigm. Delicate as it is, that slight susurration is powerful enough to shatter the walls of every scientific institution on Earth.

Is the sound made by a ghost in your machine? Or is it a quirk of perception? A tiny taste of madness that leaves us wanting more? In the end, it doesn't matter. Whether the spirits are there or not, our embrace is what pushes our institutions into a state of weightless ephemera.

I did this. I listened. On the first day, when I played it back, I heard nothing. On the second day, nothing.

If I did hear something, would I trust the science? Or myself?

The third through the ninth days, nothing.

On the tenth day, I huddled in the silence, ready to listen with my eyes closed so that I would pick up the slightest whisper. I knew that, according to an interpretation of science, there was nobody there, nothing but hisses. But still. But, still. But if. I pushed play.

ACKNOWLEDGMENTS

I am, first and foremost, so very grateful to the readers who picked up this book, and who have built a wonderful community of thoughtful nonfiction readers and writers. If you enjoyed reading *The Ghost Lab* in the slightest, you will join me in thanking the many people whose efforts and kindnesses allowed it to happen. At PublicAffairs, this includes my editors Ben Adams (who helped hatch the concept) and Anupama Roy-Chaudhury (who helped rear it to maturity). The endlessly effervescent Melissa Veronesi somehow makes the production train run on time without resorting to even a hint of despotism, while Joseph Gunther's dedicated copyediting pulled roughly one zillion burrs from my draft's rough and unrefined coat. Elisa Rivlin brought ample humanity to her legal review, and publicist extraordinaire Brooke Parsons continues to dazzle on all fronts. I am hopeful that, in seventy or so years, we can all get back together to publish something substantial. Or, more likely, ephemeral. Thank you to the ghosts, the cryptids, the extraterrestrials, and the spirits who continue to evoke strange reactions from all of us. And an enormous debt of gratitude goes to all those who spent hours of their busy lives talking openly with me about their personal experiences and beliefs. This applies particularly to KRI's Andy Kitt, Beau Esby, Valerie Roy, and Antje Bourdages who all, remember, spoke on the record only because they trusted that the book would fairly represent their viewpoints. I hope their expectations have borne out. They have earned both my gratitude

Acknowledgments

and respect. Thanks to Sadie Vincent for her sparkling research and formatting help. Thank you to my niece Grace and my sister Jen, Top Five people who selflessly gave their time so that I could write when I needed to. And, of course, to my beautiful and loving wife, Kimberly, who (to paraphrase Homer Simpson) has not once, in twenty years, ever eaten the last slice of pizza. You're my everything.

BIBLIOGRAPHY

Alchetron. "Oxalobacter formigenes." January 1, 2024. https://alchetron.com/Oxalobacter-formigenes.

Alper, Becka A., Michael Rotolo, Patricia Tevington, Justin Nortey, and Asta Kallo. *Spirituality Among Americans.* Pew Research Center, December 7, 2023. www.pewresearch.org/religion/2023/12/07/spirituality-among-americans/.

Angier, Natalie. "The Moon Comes Around Again." *New York Times*, September 7, 2014. www.nytimes.com/2014/09/09/science/revisiting-the-moon.html.

Anthes, Emily. "The Animal Translators." *New York Times*, August 30, 2022. www.nytimes.com/2022/08/30/science/translators-animals-naked-mole-rats.html.

Arthur Conan Doyle Encyclopedia. "Lectures at the International Spiritualist Congress of Paris 1925." N.d. www.arthur-conan-doyle.com/index.php?title=Lectures_at_the_International_Spiritualist_Congress_of_Paris_1925.

Associated Press. "AG Takes Drug-Raid Team to Task over Fatal Greenland Shootout." *Portland Press Herald*, December 14, 2012. www.pressherald.com/2012/12/14/ag-takes-drug-raid-team-to-task-over-fatal-greenland-shoot-out-_2012-12-15/.

Associated Press. "Documents: NH Man Called Friend to Admit Shootings." WPTZ Burlington, April 23, 2012. www.mynbc5.com/article/documents-nh-man-called-friend-to-admit-shootings/3303066.

Associated Press. "UNH Lays Off 20 Staffers." WPTZ Burlington, March 20, 2012. www.mynbc5.com/article/unh-lays-off-20-staffers/3302666.

Beaning, David. *Law School Involvement in Community Development: A Study of Current Initiatives and Approaches.* US Department of Housing and Urban Development, 1998. www.huduser.gov/portal/publications/pdf/LawSchoolInvolvement.pdf.

Bibliography

Beer, Matt. "Feds Curious About High-Tech Conclave on Extraterrestrials." *SFgate*, October 8, 1999. www.sfgate.com/business/article/Feds-curious-about-high-tech-conclave-on-3064331.php.

Beimler, Nik. "'Remember, Celebrate and Pray.'" Seacoast Online, May 28, 2016. www.seacoastonline.com/story/news/2016/05/29/remember-celebrate-pray/28412725007/.

Beischel, Julie. "Spontaneous, Facilitated, Assisted, and Requested After-Death Communication Experiences and Their Impact on Grief." *Threshold: Journal of Interdisciplinary Consciousness Studies* 3, no. 1 (May 2019). www.researchgate.net/publication/334330476_Spontaneous_Facilitated_Assisted_and_Requested_After-Death_Communication_Experiences_and_their_Impact_on_Grief_Peer-reviewed_referenced_commentary.

Bernard, Chris. "Moving Moments." Seacoast Online, January 31, 2011. www.seacoastonline.com/story/news/2004/10/08/moving-moments/50247008007/.

Bouchard, Kleinman & Wright Law Firm. "NH Supreme Court—No Insurance Coverage for Mom of Police Shooter." BKW Law, November 13, 2014. https://bkwlawyers.com/new-hampshire-supreme-court-finds-no-insurance-coverage-for-mother-of-police-shooter/.

Brenan, Megan. "Americans More Critical of US Criminal Justice System." Gallup, November 16, 2023. https://news.gallup.com/poll/544439/americans-critical-criminal-justice-system.aspx.

Brittle, Gerald. *The Demonologist: The Extraordinary Career of Ed and Lorraine Warren*. New York: Graymalkin Media, 2013.

Brode, Noah. "US Belief in Sasquatch Has Risen Since 2020." CivicScience, August 2, 2022. https://civicscience.com/u-s-belief-in-sasquatch-has-risen-since-2020/.

Brown, Janice. "New Hampshire Senator, Legislator, Farmer, Livestock Breeder and Mail Carrier: Doris M. Spollett of Hampstead." *Cow Hampshire: New Hampshire's History Blog*, December 27, 2019. www.cowhampshireblog.com/2019/12/27/new-hampshire-senator-legislator-farmer-livestock-breeder-and-mail-carrier-doris-m-spollett-of-hampstead/.

Burton, Scott. "Body Armor Expert Details Limitations of Bullet-Proof Vests." Body Armor News, n.d. www.bodyarmornews.com/body-armor-expert-details-limitations-of-bullet-proof-vests/.

Button Factory Artists' Studios. "The Story of Morley." The Button Factory, n.d. www.buttonfactorystudios.com/images/morely-a.jpg.

Byrne, Brendan, ed. "Simon Parkes, British City Councilman, Claims Alien Contact and Child." Value Walk, September 19, 2021. www.valuewalk.com/simon-parkes-claims-alien-contact-child/.

Cabral, Lídia. "Decentralisation in Africa: Scope, Motivations and Impact on Service Delivery and Poverty." Working paper, Future Agricultures Consortium, March 2011. https://assets.publishing.service.gov.uk/media/57a08ae740f0b652dd000 98c/FAC_Working_Paper_No20.pdf.

Callie. "Churchill's WW2 Speech to the Nation October 1939." *WW2 Memories* (blog), September 24, 2011. https://ww2memories.wordpress.com/2011/09/24 /churchills-ww2-speech-to-the-nation-october-1939/.

Carlson, Andy. *State Higher Education Finance FY 2011*. State Higher Education Executive Officers, 2012. https://sheeo.org/wp-content/uploads/2019/03/SHEF _FY11-2.pdf.

Carpenter, Alexa. "UNH Cuts 17 Lecturer Positions." *Saint Anselm Crier*, March 2, 2018. https://criernewsroom.com/news/2018/03/02/unh-cuts-17-lecturer-potions/.

CBS Boston and Associated Press. "Woman Killed in Greenland Shooting Was Worried About Mutrie." CBS News, April 13, 2012. www.cbsnews.com/boston /news/nh-police-shooting-suspect-was-a-former-firefighter/.

Center for Community, Culture and Change. "The Center." N.d., archived January 31, 2009. https://web.archive.org/web/20090131134450/http://thecenter4ccc .org/Site/Home.html.

Chen, Elaine. "Can You Understand Bird? Test Your Recognition of Calls and Songs." *New York Times*, July 21, 2023. www.nytimes.com/interactive/2023/07/21/science /bird-calls-songs-quiz.html.

Chiang, Eva. "Distrust in Our Institutions Is Real, but We Can Rebuild Trust." *The Catalyst*, no. 22 (Winter 2022). www.bushcenter.org/catalyst/restoring-trust -in-institutions/chiang-distrust-in-institutions-is-real-how-to-rebuild-trust.

Christy, Michael. "The Reincarnations of General Patton." *Together We Served* (blog), n.d. https://blog.togetherweserved.com/the-reincarnations-of-general-patton/.

Cohen, Rebecca. "Breaking Down Public Trust." Gerald R. Ford School of Public Policy, June 10, 2021. https://fordschool.umich.edu/news/2021/rebuilding-trust -in-government-democracy.

Coleman, Loren. "Comic Book Artist Michael Mitchell, 53, Dies Suddenly." CryptoZooNews, September 10, 2019. www.cryptozoonews.com/mitchel-m-obit/.

Commonwealth of Massachusetts. "Mary J. Whalen, Administratrix, vs. Patrick B. Mutrie." Mass Cases, n.d. http://masscases.com/cases/sjc/247/247mass316 html.

Conklin, Susan. "Rules & FAQ." Rolling Hills Asylum, n.d. www.rollinghillsasylum.com/get-started/rules-and-faq.

Cook, Robert. "Civil Suit: Beverly Mutrie Provided Gun That Killed Greenland Police Chief." Patch, November 18, 2012. https://patch.com/new-hampshire/portsmouth-nh/civil-suit-beverly-mutrie-provided-gun-that-killed-grd6a9b60e00.

Costandi, Moheb. "What Happens to Our Bodies After We Die." BBC News, May 8, 2015. www.bbc.com/future/article/20150508-what-happens-after-we-die.

Coulthart, Ross. "Rep. Burchett: Believing UFOs Are in the Bible Is Not Anti-Christian." Produced by Andy Gipson. *Reality Check* (podcast), April 3, 2024. YouTube, 45:23. www.youtube.com/watch?v=c5iPtsQy2cg.

Cresta, Joey. "AG Probe: Mutrie Had Lengthy Violent History." Seacoast Online, December 17, 2012. www.seacoastonline.com/story/news/local/portsmouth-herald/2012/12/17/ag-probe-mutrie-had-lengthy/49215122007/.

Crimesider Staff. "In UNH Murder Case, Why Didn't Witnesses Call Police?" CBS News, February 17, 2015. www.cbsnews.com/news/in-lizzi-marriott-university-of-new-hampshire-murder-case-why-didnt-witnesses-call-police/.

Cromell, Chris. *Ghost Hunting: A Biblical Perspective*. Ebook, 2010. www.amazon.com/Ghost-Hunting-Perspective-Chris-Cromell-ebook/dp/B0042JTN3K.

Cromie, William J. "Alien Abduction Claims Examined." Harvard Gazette, February 20, 2003. https://news.harvard.edu/gazette/story/2003/02/alien-abduction-claims-examined-2/.

Cullum, Tommy. "Granite Sky with Mike Stevens." *Let's Get Freaky* (podcast), April 9, 2024. YouTube, 1:04:48. www.youtube.com/watch?v=qWOPG5SuDMQ.

Cunningham, Able. "Bigfoot Witness Accounts." 603 Bigfoot, n.d. https://603bigfoot.com/witness-accounts.

"Colliers International Sells Two Properties Totaling $4.92 Million—Including the Former Mcintosh College Main Campus." *New England Real Estate Journal*, June 1, 2018. https://nerej.com/colliers-international-sells-two-properties-total ing-4-92-million.

Daily News. "Coyle Talks about Rolling Hills Asylum." February 16, 2016. www.thedailynewsonline.com/news/coyle-talks-about-rolling-hills-asylum/article_6da95773-a16c-5f0e-b1ac-4db0ed4a31c2.html.

Darwin, C. R. *The Expression of the Emotions in Man and Animals*. London: John Murray, 1872.

Dean, Cornelia. "Eager to Tell the Stories of Science, a Biologist Evolves." *New York Times*, April 11, 2006. www.nytimes.com/2006/04/11/science/sciencespecial2/eager-to-tell-the-stories-of-science-a-biologist.html.

Bibliography

Dillon, Michele, and Shannon H. Rogers. *New Hampshire Civic Health Index 2009.* Durham, NH: Carsey Institute, 2009. https://dx.doi.org/10.34051/p/2020.83.

Dinan, Elizabeth. "Officers Wounded by Mutrie Lose Civil Suit Against Mother." Seacoast Online, July 11, 2014. www.seacoastonline.com/story/news/local/portsmouth-herald/2014/07/11/officers-wounded-by-mutrie-lose/36464991007/.

Dinan, Elizabeth. "Police Shooter's Mom Had 'Armory,' Says Lawyer." Seacoast Online, November 15, 2012. www.seacoastonline.com/story/news/local/hampton-union/2012/11/16/police-shooter-s-mom-had/49274062007/.

Diocese of Manchester. "Merger of St. Charles Borromeo Parish, St. Joseph Parish, and St. Mary Parish in Dover 07.04.09." Catholic Church in New Hampshire, 2009. www.catholicnh.org/assets/Documents/Parishes/Decrees/Merger/GrovetonNStratfordDecree.pdf.

Diocese of Manchester. "St. Patrick Church." Catholic Church in New Hampshire, accessed July 7, 2024. https://directory.catholicnh.org/churchsearch/view/4.

Do Good Institute. "New Research Report Examines Recent Declines in Giving and Volunteering in America." University of Maryland School of Public Policy, November 1, 2023. https://dogood.umd.edu/news/new-research-report-examines-recent-declines-giving-and-volunteering-america.

Douthat, Ross. "Be Open to Spiritual Experience. Also, Be Really Careful." *New York Times*, February 1, 2023. www.nytimes.com/2023/02/01/opinion/american-religion-spirituality.html.

Douthat, Ross. "Flying Saucers and Other Fairy Tales." *New York Times*, December 23, 2017. www.nytimes.com/2017/12/23/opinion/alien-encounters-christmas-ufo.html.

Douthat, Ross. "What the News and the Pews Have in Common." *New York Times*, February 16, 2024. www.nytimes.com/2024/02/16/opinion/religion-internet.html.

Duncan, Grant. "How to Restore Trust in Governments and Institutions." The Conversation, November 9, 2018. https://theconversation.com/how-to-restore-trust-in-governments-and-institutions-106547.

Enders, Adam, Christina Farhart, Joanne Miller, Joseph Uscinski, Kyle Saunders, and Hugo Drochon. "Are Republicans and Conservatives More Likely to Believe Conspiracy Theories?" *Political Behavior* 45, no. 4 (2023). https://doi.org/10.1007/s11109-022-09812-3.

Esby, Troy T. "What's Up?" The Center, n.d. https://web.archive.org/web/20090930192830/http://www.thecenter4ccc.org/Site/Home.html.

EurekAlert! "Links Between Paranormal Beliefs and Cognitive Function Described by 40 Years of Research." May 4, 2022. www.eurekalert.org/news-releases/950987.

Fensom, Gail. "Coming to Know Reflective Practice: An Ethnography of Novice University Teachers." Phd diss., University of New Hampshire, 2007. Page 317. https://scholars.unh.edu/dissertation/371.

Fink, Richard W. "The Commercialization of the Afterlife: Spiritualism's Supernatural Economy, 1848–1900." Master's thesis, Temple University, 2010. http://dx.doi.org/10.34944/dspace/1196.

Fleck, Anna. "Is US Belief in the Afterlife Dying?" Statista, August 2, 2023. www.statista.com/chart/30526/share-of-us-respondents-who-believe-in-the-afterlife/.

Foster, Joanna M. "Warming Ski Slopes, Shriveled Revenues." *New York Times,* December 7, 2012. https://archive.nytimes.com/green.blogs.nytimes.com/2012/12/07/warming-slopes-shriveling-revenues/.

Foster's Daily Democrat. "4 Wounded NH Officers Released from Hospital." Officer, April 27, 2012. www.officer.com/tactical/firearms/news/10706259/4-wounded-nh-officers-released-from-hospital.

Fountain, Henry. "Bees like a Warm Drink Now and Again, Too." *New York Times,* August 15, 2006. www.nytimes.com/2006/08/15/science/15observ.html.

Fountain, Henry. "Observatory." *New York Times,* August 17, 2004. www.nytimes.com/2004/08/17/science/observatory.html.

Freund, Alexander. "The Science of Dying." Deutsche Welle, April 17, 2019. www.dw.com/en/the-science-of-dying/a-48372592.

Gaither, Milton. "State Homeschool Enrollment Data Trends, 2014." International Center for Home Education Research Reviews, December 24, 2014. https://icher.org/blog/?p=1459.

Gallup Polls. "Evolution, Creationism, Intelligent Design." Gallup, March 22, 2024. https://news.gallup.com/poll/21814/evolution-creationism-intelligent-design.aspx.

Gallup Polls. "Party Affiliation." Gallup, 2024. https://news.gallup.com/poll/15370/party-affiliation.aspx.

Ganley, Rick, and Mary McIntyre. "UNH Releases Five More Lecturers; Japanese Program Eliminated." New Hampshire Public Radio, November 27, 2019. www.nhpr.org/post/unh-releases-five-more-lecturers-japanese-program-eliminated.

Gmelch, George. "Baseball Magic." *Trans-action* 8, no. 8 (June 1971). www.meissinger.com/uploads/3/4/9/1/34919185/gmelch_baseball_magic.pdf.

Goldthwait, James Walter, Lawrence Goldthwait, and Richard Parker Goldthwait. *The Geology of New Hampshire: Part I—Surficial Geology.* Concord, NH: New Hampshire State Planning and Development Commission, 1951. www.des.nh.gov/sites/g/files/ehbemt341/files/documents/geo-nhx-250000-sbsm-geologynhparti-final.pdf.

Bibliography

Greek Obituary. "Leonidas P. Chetsas of Newburyport, Massachusetts: 1928–2017." April 2017. www.greekobituary.net/obituary/leonidas-chetsas.

Haddadin, Jim. "3,000-Plus Page Report Provides Insight into Greenland Tragedy." *Foster's Daily Democrat*, April 12, 2013. www.fosters.com/story/news/2012/12/15/3-000-plus-page-report/48959136007/.

Haddadin, Jim. "Police: Mutrie Confessed to Shooting Police in Phone Call to Friend." *Foster's Daily Democrat*, April 12, 2013. www.fosters.com/story/news/crime/2012/04/23/police-mutrie-confessed-to-shooting/48957933007/.

Hale, Elizabeth R. "Sometimes the Cure Is Part of the Disease: Possession, Abduction, and Spirituality in Mental Health Treatments." Master's thesis, University of California, Santa Barbara, 2022. https://escholarship.org/uc/item/1k54s35v.

Hanson, Brittany. "Shifting Systems of Faith Fuel a Rise in Paranormal Belief." Chapman Newsroom, October 23, 2019. https://news.chapman.edu/2019/10/23/cultural-research-project-points-to-a-rise-in-paranormal-belief/.

Hauswitch Home + Healing. "Witch the Vote: Black Lives Matter." *HausWitch* (blog), June 2, 2020. https://hauswitchstore.com/blogs/community/witch-the-vote-black-lives-matter.

Henderson, Shawn. "The Paranormal Experience Ghosts, Science, and the Number 42." *Spirit Radio: The Paranormal Experience*. Radio broadcast, Portsmouth, NH, January 23, 2016. www.youtube.com/watch?v=QiMkDnP_H18.

Hernandez, Nestor, and Paul Hemez. "Some Demographic and Economic Characteristics of Male and Female Same-Sex Couples Differed." US Census Bureau, November 8, 2023. www.census.gov/library/stories/2023/11/same-sex-couple-diversity.html.

Higginbottom, Justin. "The 'Psychonauts' Training to Explore Another Dimension." *New Republic*, January 4, 2023. https://newrepublic.com/article/169525/psychonauts-training-psychedelics-dmt-extended-state.

History of Hypnosis. "Hypnosis in the 19th Century." N.d. www.historyofhypnosis.org/19th-century.html.

Hunt, Donald. "Oritha M. Miller, One of the First Black Female Marines, Dies at 89." *Philadelphia Tribune*, October 13, 2021. www.phillytrib.com/obituaries/oritha-m-miller-one-of-the-first-black-female-marines-dies-at-89/article_7487f969-a039-5cf1-a7c8-8122352e5634.html.

Hutchings, Emily Grant. *Jap Herron: A Novel Written from the Ouija Board*. New York: Mitchell Kennerley, 1917. www.gutenberg.org/cache/epub/33048/pg33048-images.html.

Bibliography

Ingmire, Bruce E. "The Morely Button Factory." Portsmouth Athenaeum, n.d. https://ppolinks.com/athenaeum/vertical_file_0102_01.pdf.

Jinks, Tony. *An Introduction to the Psychology of Paranormal Belief and Experience.* Jefferson, NC: McFarland, 2011.

Johnson, Keith. *Paranormal Realities.* Summer Wind Press, 2009.

Jones, Jeffrey M. "Independent Party ID Tied for High; Democratic ID at New Low." Gallup, January 12, 2024. https://news.gallup.com/poll/548459/independent-party-tied-high-democratic-new-low.aspx.

Kates, Sean, Jonathan Ladd, and Joshua A. Tucker. "How Americans' Confidence in Technology Firms Has Dropped: Evidence from the Second Wave of the American Institutional Confidence Poll." Brookings, June 14, 2023. www.brookings.edu/articles/how-americans-confidence-in-technology-firms-has-dropped-evidence-from-the-second-wave-of-the-american-institutional-confidence-poll/.

Kihlstrom, John F., Heather A. Brenneman, Donna D. Pistole, and Ronald E. Shor. "Hypnosis as a Retrieval Cue in Posthypnotic Amnesia." *Journal of Abnormal Psychology* 94, no. 3 (1985). https://doi.org/10.1037/0021-843X.94.3.264.

King, David C., and Zachary Karabell. *The Generation of Trust: Public Confidence in the US Military Since Vietnam.* American Enterprise Institute, 2003. www.aei.org/wp-content/uploads/2014/07/-the-generation-of-trust_162530927872.pdf?x85095

Kitt, Andrew J. *Ghosts, Science, and the Number 42: The Success of Ghost Hunting and the Failure of Skepticism.* Portsmouth, NH: Walking Elk Publications, 2012.

Kreider, Rose M. "Intermarriage: Profiles of the Most Common Interracial Combinations Using 1990 Census Data." US Census Bureau, October 2000. www.census.gov/library/working-papers/2000/demo/kreider-00.html.

Lazar, Kay. "Michael Maloney Recalled for Kindness, Love of His Work." *Boston Globe*, April 14, 2012. www.bostonglobe.com/metro/2012/04/13/greenland-police-chief-michael-maloney-felled-nighttime-barrage-recalled-cop-cop/33IYYw2UOfJb2ESZhETskN/story.html.

Leapin' Lizards Gift and Holistic Center. "Medium Isabeau Esby." Facebook, 2014. www.facebook.com/events/leapin-lizards/medium-isabeau-esby/149903631 7002954/.

Lenahan, Ian. "'Missed More Than You Know': Chief Michael Maloney's Death Stirs Emotions 10 Years Later." Seacoast Online, April 12, 2022. www.seacoastonline.com/story/news/local/2022/04/12/greenland-nh-police-chief-michael-maloney-died-decade-ago/7287356001/.

Bibliography

Leroux, Gaston. *The Phantom of the Opera*. 1910. www.gutenberg.org/files/175/175-h/175-h.htm.

Lessard, Ryan, and Kelly Sennott. "How Hesser Failed." *Hippo*, December 17, 2015. https://web.archive.org/web/20191226020210/http://hippopress.com/read-article/how-hesser-failed.

Lilith Cult. "The Truth About Parasite Spirits." *Lilith Cult* (blog), September 3, 2022. https://lilithcult.com/blogs/get-to-know-more-about-demonology-and-black-magic/parasite-spirits.

Malinowski, Bronislaw. "53. Fishing in the Trobriand Islands." *Man* 18 (June 1918). https://doi.org/10.2307/2788612.

Mallory, Bruce L., and Quixada. Moore-Vissing. *2012 New Hampshire Civic Health Index*. Durham, NH: Carsey Institute, 2013. https://dx.doi.org/10.34051/p/2020.193.

Mallory, Bruce L., and Quixada Moore-Vissing. *2020 New Hampshire Civic Health Index*. Durham, NH: Carsey School of Public Policy, 2013. https://dx.doi.org/10.34051/p/2021.17.

Mandelstam, Joel. "Du Chaillu's Stuffed Gorillas and the Savants from the British Museum." *Notes and Records of the Royal Society of London* 48, no. 2 (1994): 227–245. Accessed April 26, 2024. www.jstor.org/stable/532165.

McCarthy, Caroline. "Silicon Valley Has a Problem with Conservatives. But Not the Political Kind." *Vox*, June 12, 2018. www.vox.com/first-person/2018/6/12/17443134/silicon-valley-conservatives-religion-atheism-james-damore.

McCartin, Jeanne. "Closing of UNH's Museum of Art 'Leaves Giant Hole for Community.'" *NH Business Review*, January 23, 2024. www.nhbr.com/closing-of-unhs-museum-of-art-leaves-giant-hole-for-community/.

Mervosh, Sarah. "Teachers Are Missing More School, and There Are Too Few Substitutes." *New York Times*, February 19, 2024. www.nytimes.com/2024/02/19/us/teacher-absences-substitute-shortage.html.

Mervosh, Sarah, and Francesca Paris. "Why School Absences Have 'Exploded' Almost Everywhere." *New York Times*, March 29, 2024. www.nytimes.com/interactive/2024/03/29/us/chronic-absences.html.

Microbiota Group of Iran. "Adlercreutzia." N.d. https://microbiomology.org/microbe/adlercreutzia/.

Moore, Wendy. "Benjamin Franklin and the Glass Armonica." Benjamin Franklin House, n.d. https://benjaminfranklinhouse.org/benjamin-franklin-and-the-glass-armonica/.

Bibliography

Mottin, Don. "Advanced Hypnosis Training." Mottin & Johnson Institute of Hypnosis, n.d. https://donmottin.com/.

Neuman, Scott. "The Faithful See Both Crisis and Opportunity as Churches Close Across the Country." NPR, May 17, 2023. www.npr.org/2023/05/17/1175452002/church-closings-religious-affiliation.

New Hampshire. "Chairman of Psychology Department Found Dead." February 2, 1982. https://scholars.unh.edu/cgi/viewcontent.cgi?article=3766&context=tnh_archive.

New Hampshire Department of Education. "Closed Colleges and Universities." N.d. www.education.nh.gov/who-we-are/division-of-educator-support-and-higher-education/closed-colleges-universities.

New Hampshire Department of Education. *State Totals Ten Years Public and Private Fall Enrollments: 2014–2015 Through 2023–2024.* February 23, 2024. https://my.doe.nh.gov/iPlatform/Report/Report?path=%2FBDMQ%2FiPlatform%20Reports%2FEnrollment%20Data%2FState%20Totals%2FState%20Totals%20Ten%20Years%20Public%20and%20Private%20Fall%20Enrollments&name=State%20Totals%20Ten%20Years%20Public%20and%20Private%20Fall%20Enrollments&categoryName=State%20Totals&categoryId=12#.

New Hampshire PBS. "NHPTV Restructures Due to Elimination of $2.7 Million State Appropriation." June 1, 2011. https://nhpbs.org/pressroom/release_detail.asp?hp_id=1259.

New Hampshire Public Radio. "Facing Annual Loss, UNH Law Says Deficit Is Part of Smart Strategy." October 28, 2019. www.nhpr.org/nh-news/2019-10-28/facing-annual-loss-unh-law-says-deficit-is-part-of-smart-strategy.

NH Business Review Staff. "Reorganized UNH Cooperative Extension Tries to Cope with Cuts." *NH Business Review*, January 5, 2012. www.nhbr.com/reorganized-unh-cooperative-extension-tries-to-cope-with-cuts/.

NH Law Enforcement Officers' Memorial Association. "Chief Michael Maloney." Accessed July 7, 2024. https://nhlawenforcementmemorial.com/portfolio-item/michael-maloney/.

Orth, Taylor. "A Growing Share of Americans Believe Aliens Are Responsible for Ufos." YouGov, October 5, 2022. https://today.yougov.com/topics/technology/articles-reports/2022/10/04/more-half-americans-believe-aliens-probably-exist.

Ottman, Noora A. "Host Immunostimulation and Substrate Utilization of the Gut Symbiont Akkermansia muciniphila." PhD diss., Wageningen University and Research, 2015. Figure 3. www.researchgate.net/figure/Electron-microscope-images-of-A-muciniphila-Images-by-Justus-Reunanen-University-of_fig3_283273102.

Bibliography

Overbye, Dennis. "Deep in the Cosmic Forest, a Black Hole Goldilocks Might Like." *New York Times,* May 6, 2020. www.nytimes.com/2020/05/06/science/black-hole-intermediate.html.

Partners for Sacred Places. "Economic Halo Effect." N.d. https://sacredplaces.org/info/publications/halo-studies/.

Pew Research Center. "Public Trust in Government: 1958–2024." June 24, 2024. www.pewresearch.org/politics/2024/06/24/public-trust-in-government-1958-2024/.

Pretorius, Christina. "Where Do New Hampshire Students Go to School? 3 Key Takeaways on K–12 School Enrollment." Reaching Higher NH, January 5, 2023. https://reachinghighernh.org/2023/01/05/where-do-new-hampshire-students-go-to-school-3-key-takeaways-on-k-12-school-enrollment/.

Psychology.org. "Demand for Psychology in New Hampshire." Accessed August 5, 2022. www.psychology.org/online-degrees/new-hampshire/#demand.

Psychology.org. "Psychology in New Hampshire: Learn About Becoming a Psychologist in NH." Accessed August 5, 2022. www.psychology.org/online-degrees/new-hampshire/.

Radford, Benjamin. "Ghost Hunters in the Dark." *Skeptical Inquirer.* Center for Inquiry, 2017. https://skepticalinquirer.org/2017/01/ghost-hunters-in-the-dark/.

Ramer, Holly. "NH Panel to Review Police Shooting." *Telegram & Gazette*, May 2, 2012. www.telegram.com/story/news/local/north/2012/05/02/nh-panel-to-review-police/49650595007/.

Rate My Professors. "Andrew Kitt at University of New Hampshire." N.d. www.ratemyprofessors.com/professor/2071398.

Raynor, Shane. "Christians and Ghost Hunting." Ministry Matters, October 12, 2012. www.ministrymatters.com/all/entry/3323/christians-and-ghost-hunting.

Real Holistic Doc. "PEMF Device." N.d. https://realholisticdoc.com/pemf-device/.

Rendle, Gil. "Restoring Trust in the Institutional Church Requires Allegiance to Our True Purpose—Lewis Center for Church Leadership." Lewis Center for Church Leadership, February 6, 2024. www.churchleadership.com/leading-ideas/restoring-trust-in-the-institutional-church-requires-allegiance-to-our-true-purpose/.

Rimkunas, Barbara. "George Dearborn and the Spiritualists." Exeter Historical Society, October 22, 2020. www.exeterhistory.org/historically-speaking/2020/10/22/george-dearborn-and-the-spiritualists.

Robinson, J. Dennis. "The Incident at Exeter High." Seacoast NH, November 2000. www.seacoastnh.com/the-incident-at-exeter-high/?start=1.

Bibliography

Robson, David. "Psychology: The Truth About the Paranormal." *BBC News*, October 31, 2014. www.bbc.com/future/article/20141030-the-truth-about-the-paranormal.

Rousseau, Morgan. "NH Ranked the Worst State for Porch Pirates: Report." Boston .com, June 4, 2023. www.boston.com/news/local-news/2023/06/04/n-h-ranked-the-worst-state-for-porch-pirates-report/.

Routledge, Clay. "Don't Believe in God? Maybe You'll Try UFOs." *New York Times*, July 21, 2017. www.nytimes.com/2017/07/21/opinion/sunday/dont-believe-in-god-maybe-youll-try-ufos.html.

Russonello, Giovanni. "QAnon Now As Popular in US As Some Major Religions, Poll Suggests." *New York Times*, August 12, 2021. www.nytimes.com/2021/05/27/us/politics/qanon-republicans-trump.html.

Sakellarios, Stephen. "ssake1." YouTube account, n.d. www.youtube.com/@ssake1_IAL_Research/videos.

Scarborough, Dorothy. *Humorous Ghost Stories*. New York and London: Knickerbocker Press, 1921. www.gutenberg.org/files/26950/26950-h/26950-h.htm.

Schwartz, John. "The Crew of STS-123." *New York Times*, March 10, 2008. www.nytimes.com/2008/03/10/science/space/10crew.html.

Shah, Sonal, and Hollie Russon Gilman. "Rebuilding Trust in American Institutions." *Stanford Social Innovation Review*, January 27, 2021. https://doi.org/10.48558/RDSQ-Z185.

Singer, Barry, and Victor A. Benassi. "Occult Beliefs: Media Distortions, Social Uncertainty, and Deficiencies of Human Reasoning Seem to Be at the Basis of Occult Beliefs." *American Scientist* 69, no. 1 (1981). www.jstor.org/stable/27850247.

Sletten, Phil. "Public Higher Education Funding in New Hampshire Trails All Other States Despite Recent Increases." *New Hampshire Fiscal Policy Institute* (blog), June 14, 2022. https://nhfpi.org/blog/public-higher-education-funding-in-new-hampshire-trails-all-other-states-despite-recent-increases/.

Smith, Kerri. "PastCast: Gorillas, Man-Eating Monsters?" *Nature PastCast* (podcast), June 28, 2019. https://doi.org/10.1038/d41586-019-01883-3.

Smith, Tom W. "Trends in Confidence in Institutions, 1973–2006." *Social Trends in American Life: Findings from the General Social Survey since 1972* (2012). https://gss.norc.org/Documents/reports/social-change-reports/SC54%20Trends%20in%20Confidence%20in%20Institutions.pdf.

Spadaro, Giuliana, Katharina Gangl, Jan-Willem Van Prooijen, Paul A. M. Van Lange, and Cristina O. Mosso. "Enhancing Feelings of Security: How Institutional Trust Promotes Interpersonal Trust." *PLOS One* 15, no. 9 (September 11, 2020). https://journals.plos.org/plosone/article?id=10.1371/journal.pone.0237934.

Stroebe, Margaret, Wolfgang Stroebe, and Georgios Abakoumkin. "The Broken Heart: Suicidal Ideation in Bereavement." *American Journal of Psychiatry* 162, no. 11 (2005). https://ajp.psychiatryonline.org/doi/pdf/10.1176/appi.ajp.162.11.2178.

Stuart, Nancy Rubin. *The Reluctant Spiritualist: The Life of Maggie Fox*. Orlando, FL: Harcourt, 2005.

Sullivan-Bissett, Ema. "Is It Normal to Believe You Have Been Abducted by Aliens?" University of Birmingham. N.d. www.birmingham.ac.uk/research/perspective/abducted-by-aliens.

Thomas Institute of Hypnosis. "Hypnotherapists/Instructors of Thomas Institute of Hypnosis." N.d. https://ourmagic1.tripod.com/your_hypnotherapist.html.

"Three Forms of Thought; M. M. Mangassarian Addresses the Society for Ethical Culture at Carnegie Music Hall; Unrest of the Human Mind; Theosophy, Spiritualism, and Christian Science Discussed—the Theory of Reaction a Fallacy—Ineffectiveness of the Spiritualistic Idea." *New York Times*, November 29, 1897. www.nytimes.com/1897/11/29/archives/three-forms-of-thought-mm-mangassarian-addresses-the-society-for.html.

Thumma, Scott. "Twenty Years of Congregational Change: The 2020 Faith Communities Today Overview." Hartford Institute for Religion Research, 2021. https://faithcommunitiestoday.org/wp-content/uploads/2021/10/Faith-Communities-Today-2020-Summary-Report.pdf.

Tim Gaudreau Studios. "Portfolio." N.d. https://timgaudreau.com/.

Timmins, Annmarie. "NH Has Lost 11 Maternity Wards in 20 Years. Frisbie Memorial Is the Latest." New Hampshire Public Radio, March 9, 2023. www.nhpr.org/health/2023-03-09/nh-has-lost-11-maternity-wards-in-20-years-frisbie-memorial-is-the-latest.

Tough, Paul. "Americans Are Losing Faith in the Value of College. Whose Fault Is That?" *New York Times*, September 5, 2023. www.nytimes.com/2023/09/05/magazine/college-worth-price.html.

Twenge, Jean M., W. Keith Campbell, and Nathan T. Carter. "Declines in Trust in Others and Confidence in Institutions Among American Adults and Late Adolescents, 1972–2012." *Psychological Science* 25, no. 10 (September 9, 2014). https://doi.org/10.1177/0956797614545133.

UNH News. "UNH Prof. Mark Henn Discusses His Class 'Psychology, Critical Thinking and the Scientific Method—Exploring Paranormal Belief.'" Vimeo, 2010. https://vimeo.com/11742363.

University of New Hampshire. "Academic Buildings and Properties." N.d. https://library.unh.edu/find/archives/buildings/academic-buildings-and-properties.

University of New Hampshire. "Beyond Imagining: Capital Projects." *UNH Today*, October 19, 2019. www.unh.edu/unhtoday/2018/10/beyond-imagining-capital-projects.

University of New Hampshire. "Nanoscale Science and Engineering Center for High-Rate Nanomanufacturing." Find Scholars @ UNH, n.d. https://findscholars.unh.edu/display/grant9613.

University of New Hampshire. "Sharyn Potter." College of Liberal Arts, n.d. https://cola.unh.edu/person/sharyn-potter.

University of New Hampshire School of Law. "UNH School of Law Announces Chief Justice John Broderick as New Dean." PR Newswire, November 9, 2010. www.prnewswire.com/news-releases/unh-school-of-law-announces-chief-justice-john-broderick-as-new-dean-106995248.html.

UnivStats. "Ivy League Comparison." N.d. www.univstats.com/comparison/ncaa/ivy-league/.

Urban Institute. "Labor and Delivery Unit Closures in Rural New Hampshire." N.d. www.urban.org/policy-centers/health-policy-center/projects/labor-and-delivery-unit-closures-rural-new-hampshire.

US Bureau of Labor Statistics. "Occupational Employment and Wage Statistics, May 2019." July 6, 2020. www.bls.gov/oes/2019/may/oes259044.htm.

US Department of Justice. "288. Admissibility at Trial." *Criminal Resource Manual*. US Department of Justice Archives, n.d. www.justice.gov/archives/jm/criminal-resource-manual-288-admissibility-trial.

WBUR. "Suspect in NH Police Chief Killing Found Dead." WBUR News, April 13, 2012. www.wbur.org/news/2012/04/13/nh-shooting.

WCVB5. "Police Chief's Killer Named at Memorial Event." WCVB TV, June 4, 2013. www.wcvb.com/article/greenland-police-chief-s-killer-named-at-memorial-event/8182996#.

Wikipedia. "Center for High-Rate Nanomanufacturing," Accessed March 16, 2023. https://en.wikipedia.org/w/index.php?title=Center_for_High-rate_Nanomanufacturing&oldid=1077407503.

Wilcox, Anna. "Machine Elves or 'DMT Elves': A Journey into the DMT Spirit World." *Double Blind*, December 20, 2023. https://doubleblindmag.com/machine-elves-clockwork-elves-dmt-rick-strassman-terence-mckenna/.

Wilford, John Noble. "On Ancient Walls, a New Maya Epoch." *New York Times*, May 16, 2006. www.nytimes.com/2006/05/16/science/16maya.html.

Wise, Daniel. "Ghost Hunters' Demonic Encounters as Religious Experiences."

Journal of Gods and Monsters 3, no. 1 (2022). https://godsandmonsters-ojs-tx state.tdl.org/godsandmonsters/article/view/28.

Woetzel, Dave. "Cryptozoology & Creation Apologetics." Genesis Park. Originally published in *Creation Matters* 20, no. 4 (July–August 2015). www.genesispark.com /essays/cryptozoology-creation/.

Wynands, Anthony. "Part 1: Astrology and the 'Escapist Imagination' of H. P. Lovecraft—Graham Hancock Official Website." Graham Hancock, May 11, 2022. https://grahamhancock.com/wynandsa1/#sdendnote24anc.

Yoon, Carol Kaesuk. "From a Few Genes, Life's Myriad Shapes." *New York Times*, June 26, 2007. www.nytimes.com/2007/06/26/science/26devo.html.

Zhang, Andrew. "Tim Burchett: Americans Should Know What the Government Knows About UFOs—Politico." *Politico*, July 22, 2023. www.politico.com /news/2023/07/22/tim-burchett-ufos-00107708.

INDEX

Ackley, Helen, 228
Addams, Jane, 267
A. equolifasciens, 132
Aesop, 25
Agra schwarzeneggeri, 249
Alderton, Charles, 233
Aldrich, Thomas Bailey, 229
alien encounters
   abduction trauma, 129, 243, 271, 301, 303
   Betty and Barney Hill, 25–26, 98–100, 104, 206
   vs. jinn possession, 270–271
   media exploitation of, 302–303
   Mike Stevens, 24, 30–32, 41–42, 50–52, 105, 295
   reports of, 207
   skepticism, 99, 208
   Social Saucers, 81, 104–106, 129, 174, 242–243
   Stevens family, 105
   Travis Walton, 205–209
   treatment of abductees, 271, 302–303
   Val Lofaso, 303

aliens
   belief in, 100–103
   monetization of, 104–106, 111
   religious views on, 102, 252
   See also alien encounters; UFOs
Allison (ghost), 237, 240
Amazon, 197, 269, 279
American Cosmic (Pasulka), 102–103
Amityville Horror (film), 27
A. muciniphila, 132
animal communication, 168–169
animalcules, 4
animal magnetism, 56
animal spirits, 176–177
Arcturus, 256
Area 51, 118
Arizona State University, 70
armonicas, 56
Army Signal Corps, 9
art collections, monetization of, 299–300
artificial intelligence, 198
Ascended Beings, 233
Association of Academic Museums and Galleries (AAMG), 299–300
astral projection, 175

Index

astrology, 186
astronomy, 253–254
Atlantic News (periodical), 25
Atlantis, 16, 176
Austen, Jane, 33
A. vaderi, 249

bacteria, 132–133
Bader, Christopher, 292
banks, closures of, 37
Barotseland, 299
Batcheldor, Kenneth, 138–139
believers. *See* paranormal believers
Benassi, Victor, 70, 188–190
Berkowitz, Rita, 137
Bertrand, Eugene, 21
Betty and Barney Hill Collection, 104
Biden, Joe, 160
bidets, as units of measure, 34
Bierce, Ambrose, 195, 221
Bigfoot
   belief in, 16, 247
   calls, 246, 249
   elusiveness of, 249
   KRI hunt for, 244, 246–247, 249–250
   in New Hampshire, 246
Big Roy (ghost), 110, 115
Big Tech, 197–198, 269
bird songs, 169
black holes, 4, 253–254
Black Mountain, 251
Blackstock, Joel, 271
Bloom, Hyman, 299
Bob (spirit), 159, 166
Boo Buddy, 231
Borden, Lizzie, 232
Boston Courier (periodical), 94
Bourdages, Antje
   assaulted by ghost, 152–154
   college years, 119

domestic abuse, 120
Esby and, 224–225, 244–246
form fanatic, 213–214, 216–217, 220, 260
hearing voices, 119, 121, 259
jobs and career, 120–121, 122
Kitt and, 227–228, 244–245, 255, 260–261
KRI departure, 259–261, 275
KRI membership, 124–125, 135–136, 161
Lofaso and, 80–81, 84, 123, 124, 139–141, 151, 161, 206, 256
Mojave Desert drive, 118, 121–122
post-KRI, 285, 300–301
psychomanteum project, 178–179, 214, 216–220
Stevens and, 285–286, 289
table tipping experiment, 139–141
triple-medium walkthrough, 145–154
Tupperware event, 47–48, 119, 123, 295
See also Kitt Research Initiative (KRI)
Bourgoin, Louis, 34
Bourgoin, Mabel, 34
Boyoway (Kiriwina), 188–189
Brown, Cliff, 191–193, 196, 293
Browning, Elizabeth Barrett, 117, 279
Burakowski, Elizabeth, 242, 293
Burchett, Tim, 102, 252
Button Factory, 145–154, 295

Cactus Cats, 118
Cadillac Miller-Meteor Sentinel (1959), 39
Calcagni, Lou, 229–230, 234, 236–240
Cambridge University Press, 128
Camp Desert Rock, NV, 35
Capella, 256

Index

capitalism, 103, 233. *See also* monetization
Carey, Drew, 164–165, 166
Carringer, Joseph, 81
Carroll, Lewis, 182
Carsey School of Public Policy, 2–3, 103
Castle, Wendell, 299
CE-5. *See* Close Encounters of the Fifth Kind
Cement Monster, 118
Central Wave, 16
certification, 59, 75, 273, 292, 297
Chapman University Survey of American Fears, 292
Charlemagne, 277
Chester College, 156
Chestney, Linda, 300
Chetsas, Leon, 35
Chevrolet Tahoe Midnight Edition SUV, 39
Chief (ghost), 237
churches
    attendance, 197
    closures of, 36, 37, 269
    debunking by, 92, 93–94
    historical highway markers, 97
    views on aliens, 102, 252
    views on ghost hunting, 71
    See also religion
Churchill, Winston, 47, 98
Civilian Conservation Corps, 243–244
climate change, 242
Close Encounters of the Fifth Kind (CE-5), 252, 256–257
Coat of Arms, 30, 64, 295
Coleridge, Samuel Taylor, 77
college(s)
    closures of, 36, 37, 156–157, 268, 269
    distrust in, 267–268
    enrollment decline, 54, 268
    losing support, 54–55
    monetization of intellectual property, 103–104
    See also University of New Hampshire (UNH); *specific colleges*
Community General Hospital, 9
The Complete Idiot's Guide to Communicating with Spirits (Berkowitz), 137
Corinthian Hall, 92
Cornell Hospitality Quarterly (periodical), 233
corpse hunting, 40–43
Côte d'Ivoire, 199
Cotswold Olimpick Games, 27
Covid pandemic, 296, 300
Coyne, George, 102
Cracking Open (Esby), 82
creationism, 248–249
Creation Matters (periodical), 248
credentialing, 176–178
criminal justice system, 59, 143, 154, 197, 211–212
cryptozoology, 16, 247–249
Curtis, Karla Baker, 69, 73–75
Cuvier, Georges, 55
Cygnus X-1, 4

Daniel Webster College, 157
Daniel Webster Highway, 100
dark lightning, 253
Davenport, Reuben Briggs, 95
Davis, Glenn, 189–190
death, 127–128, 131–134, 248
deathbed visions, 9, 11
The Death Blow to Spiritualism (Davenport), 95
decomposition, 132–134, 304
DeCristofaro, Don, 236–240
deinstitutionalized states, 198–199, 269
demonology, 27–28, 29, 137, 168

Index

Dennehy, Brian, 164–165, 166
destigmatization campaigns, 272–273, 295–296
"deviant" groups, 272–273
diplopia, 7, 44
distrust
  in colleges, 267–268
  harms of, 3
  impacts of on UNH, 191, 268
  institutional, 3, 142–143, 191
  in New Hampshire, 2, 103, 268, 293
  paranormal beliefs and, 157–158, 187, 233, 263
  toll of, 268
Dixon, Franklin W., 145
dominant model of science popularization, 129–130
doomsday, 88, 158
Doublehead Mountain, 242, 248, 249
Douthat, Ross, 197–198
dowsing, 72
Doyle, Arthur Conan, 128, 141–142, 259, 291, 304
Doyle, Jean, 141
dozos (hunters), 199
Dr Pepper Museum, 232–233
Du Chaillu, Paul, 4
Durocher, Kristina L., 299–300
Dwyer, Joseph, 253, 293

Earthbound Deceased, 233
ecotourism, 231
ectoplasm, 142
"The Effects of Spatial Frequency on Local and Global Stereo Acuity" (Kitt), 207–208
Electronic Voice Phenomenon (EVP), 114
Eliot, George, 69
Elysian Space Dust, 286

EMF (electromagnetic field) detection, 231, 235, 236, 239, 298
Emmie (ghost), 110, 115
Empath Support Group, 242
Energetic Wellness Sampler, 254
Esby, Isabeau "Beau"
  background, 17
  Bourdages and, 224–225, 244–246
  buried body vision, 40–43
  communication with spirits, 18–19, 65–66
  Cracking Open, 82
  doomsday predictions, 87–88
  emotional processing of death lecture, 82
  ghost hunting advice, 304–305
  Holos experiment, 73–75, 295
  house clearing services, 47–48
  Idea Garage, 255, 264, 296
  intuitive development classes, 47–48, 79, 164, 179, 245–246
  Kitt and, 19–20, 64–65, 108, 226–227, 255, 260–265
  KRI departure, 260–265, 275
  Lofaso and, 88–89, 108, 113, 256
  marriage to Troy, 17, 107, 226, 255, 262, 264
  orb beliefs, 43, 45–46
  past skepticism, 25, 82
  personality, 88–89
  post-KRI, 295–297, 301
  press features, 25
  Rolling Hills Asylum clearing, 109–115
  SAGE classes, 245–246, 262, 264, 265, 296–297, 301
  spirit guides, 18, 110–111, 245, 296
  SPRG, 17, 19, 25
  triple-medium walkthrough, 145–154

Index

visions of Mike's alien, 31–32, 52
See also Kitt Research Initiative (KRI)
Esby, Troy, 17, 107, 226, 255, 264, 283
Esdaile, James, 53, 57
Ethan Spike stories (Whittier), 278–279
Etheric bodies, 128
evidence-based science *vs.* individual experiences, 59, 192, 200, 271–272, 304
evolution, 131, 248
Exeter Incident, 21–22, 282, 302
Exeter Terrestrial (TV show), 276
Exeter UFO Festival, 282, 283–284, 302–303
exorcism, 270
experiential tourism, 231. See also paratourism
extraterrestrials. See alien encounters; aliens; UFOs

Facebook, 208, 269
Fall River, MA, 232
Farmer, Steven, 176–178
Farmington Police Department, 203, 209
Fate (periodical), 25
Feliciano, Amy, 211
Ferguson, Sergeant, 209, 210
field biology, 247, 249
Fire in the Sky (film), 206
First Blood (film), 164
Fitzgerald, F. Scott, 285
Flock of Dodos (film), 131
Flourens, Marie Jean Pierre, 55
Forbes (periodical), 293
Ford, Gerald, 21
Ford School of Public Policy, 142–143
Fournet, Michelle, 169, 293
Fox, Kate, 92–96
Fox, Leah, 95

Fox, Maggie, 92–96
Franklin, Benjamin, 56
Franklin Pierce Law Center, 53–55, 103
Frazer, Daniel, 21–24
Free Design, 281
Freud, Sigmund, 107
Frisbie Memorial Hospital, 210
Frost, Robert, 13
Fuller, Margaret, 1, 279

Gal1 galaxy, 253
Galileo Galilei, 130, 248
Gallup, 211
Gambler's Fallacy, 64
GBPA. See Greater Boston Paranormal Associates
"General Butt Naked," 198–199
The Generation of Trust (King and Karabell), 143–144
Gerkin, Roberta, 180–182, 184–185
Ghostbusters (film), 39
"Ghostbusters case," 228
Ghost Hunters (TV show), 29, 63, 241
ghost hunting
  advice, 304–306
  author's experience with, 306
  church views on, 71
  competitive landscape of, 231
  equipment, 29, 231, 236, 298
  inclusion of mediums, 136
  Kitt's experiment, 305–306
  at KRI, 168, 222–224
  New England, 26–30
  New Hampshire, 33
  orbs, 43–46
  paratourism, 230, 231–234
  skepticism, 30, 235, 237
  television shows, 29, 63, 230, 231, 241
  vehicles, 39–40

Ghosts, Science, and the Number 42
    (Kitt), 61
ghosts and spirits
    assault by, 148–150, 152–154, 229,
        237, 295
    author's experiences with, 1–2, 306
    communication with, 13–15, 16,
        65–66, 281–282
    experiences of, 164, 208
    house clearing services, 47–48, 231
    at KRI, 167–168
    malevolent, 148–150, 152–154, 168,
        222, 237
    monetization of, 111, 115, 232–234
    parasites, 222–224, 228
    public belief in, 16, 70–71
    relationships with, 281–282
    Rolling Hills Asylum, 109–115
    Shor's, 59
    skepticism, 139
    treating with respect, 229–230
ghost towns, 118
Google, 197, 269
government, trust in, 2, 3
Goya, Francisco, 299
Granite Sky, 301
Granite State Bigfoot (Mitchell), 283
Granite State College, 43, 91, 207
Greater Boston Paranormal Associates
    (GBPA), 230, 235–240
Great Kefalonia Earthquake, 234
Green (R.E.M.), 120
Greenland Police Department, 156, 162
grief, 178–179, 216–218, 286
Guillotin, Joseph-Ignace, 56
gut biome, 128–129, 132

Hale, Eliz, 270–271
Hanks, Michele, 232, 233, 298
Hartmann, Theodore, 163

Hastings, Arthur, 178
Hattie (ghost), 110, 115
Hawes, Jason, 28–29, 62, 108
Hawthorne, Nathaniel, 251
Hays, Lola, 167n
Hellweg, Joseph, 199
Henn, Mark, 70
Hesser, Joseph, 155, 304
Hesser College, 155, 304
Heuvelmans, Bernard, 247
Hickok, Paul, 184–185
highway markers, 26, 97–98,
    100–101, 104
Hill, Barney, 25–26, 98–101, 104,
    206, 304
Hill, Betty, 25–26, 98–101, 104,
    206, 304
Holmes, Oliver Wendell, 205
Holos University Graduate Seminary,
    69–70, 73–75, 295
hospitals, closures of, 36–37, 269
H. princecharlesi, 249
Huddleston, Mark, 103, 193
Human Flourishing Lab, 198
Hunt, David, 21
Hutchings, Emily Grant, 167
hypnosis
    as an anesthetic, 57–58
    efficacy of, 271
    Kitt's, 48–50
    memories induced by, legal
        admissibility of, 53–55
    mesmerism, 55–57
    Stevens's, 50–52, 59
    taught at UNH, 55, 58
    training in, 58–59
    uses for, 58, 271, 293

Idea Garage, 255, 264, 296
The Incredible Mr. Limpet (film), 107–108

Index

individual experiences
    evidence-based science *vs.*, 59, 192, 200, 271–272, 295, 304
    of justice, 154
    Mazzaglia's, 187
    objective facts *vs.*, 227–228, 297, 302
    post-institutional futures, 195, 263
    reincarnation, 277
Institute of Transpersonal Psychology, 178
institutionalism, 97, 156, 193, 243, 267
institutions and institutional trust
    banks, 37
    Big Tech, 197–198, 269
    breakdown of, 2–3, 36–37, 103–104, 157–158
    building, 143–144
    churches, 36, 37, 269
    colleges, 36, 37, 53–55, 103–104, 155–157, 267–268, 269
    competence and, 143, 157
    criminal justice system, 59, 143, 154, 197, 211–212
    evidence-based knowledge and, 91–92, 106, 190, 192–193, 196, 243, 271–272, 291–292
    government, 2, 3
    hospitals, 36–37, 269
    law enforcement agencies, 156
    major global events and, 142–143
    media, 302
    paranormal beliefs and, 157–158, 187, 233, 263
    post-institutional, 297
    public good initiatives, 292
    rebuilding, 144, 269, 272, 291–292
    scientific institutions, 248
    Spiritualism as threat to, 142
    See also distrust
intellectual property, monetization of, 103–104

intelligent design, 248
International Spiritualist Congress, 141
intuitive development, 47–48, 79, 164, 179, 245–246
investigate, meaning of, 28
Ivy League, 33–34

Jackson, Andrew, 97
Jacobus, Cindy, 233
Jacobus, Mike, 233
Jesus, 277
Jillette, Penn, 71–72
jinn possession, 269–271
Johnson, Carl, 29
Johnson, Keith, 29
Jordan, Andrew, 253, 293
Journal Tribune (periodical), 25

K2 meters, 236–240
Kane, Elisha, 87
Kansas, flatness of, 70
Karabell, Zachary, 143–144
Keene State College, 119
Kennedy School of Government, 143
Kepler, Johannes, 130
King, David, 143–144
Kiriwina (Boyoway), 188–189
Kitt, Andy
    afterlife experiences, 49–50
    author's conversations with, 5–8, 83–84, 264–265, 293–294
    Bourdages and, 227–228, 244–245, 255, 260–261
    childhood, 11
    colorblindness, 50, 91
    diversity of knowledge, 7
    "The Effects of Spatial Frequency on Local and Global Stereo Acuity," 207–208

Index

Kitt, Andy *(continued)*
  Esby and, 19–20, 64–65, 108, 226–227, 255, 260–265
  father's death, 9–12, 13–14, 16
  ghost hunting experiment, 305–306
  Ghosts, Science, and the Number 42, 61
  at Granite State College, 43, 45, 207
  intelligence, 7, 123
  interest in the supernatural, 13–15, 16, 295
  jobs and career, 15
  KRI living space, 77–79, 215–216, 218–219
  Lofaso and, 64–65, 286–288
  past-life regressions, 48–49
  personality, 6, 108, 114, 294
  PhD candidacy, 293
  physical appearance, 7
  press features, 26
  relationship with father, 11–12
  research methodology standards, 45–46, 294–295
  romantic relationships, 15
  school years, 11–12
  skepticism, 175–176, 295
  smoking addiction, 11, 42, 49–50, 107–108, 293
  SPRG, 16–17, 25
  stamp collecting, 12, 15
  Stevens and, 26, 283–284, 289
  at UNH, 15, 91, 188, 207, 287, 293
  "What Is Paranormal, Anyhow?" 83–84
  See also Kitt Research Initiative (KRI)
Kitt, Betty, 9–10, 12, 13–14, 115
Kitt, Peter, 7–8, 9–12, 13–14, 90–91, 115, 127, 304
Kitt Research Initiative (KRI)
  alliterative event titles, 81

Bigfoot search, 244, 246–247, 249–250
building lease, 66–67, 69, 200–201
CE-5 Protocols, 252
Center for Consciousness Studies, 75, 77–84, 214, 287
closing of, 288
Code of Conduct, 214
corpse hunting, 40–43
Digeridoo Journey Circle, 81
donations, 77, 80, 81, 84, 105, 106, 124, 173, 175, 208
Empath Support Group, 242
Energetic Wellness Sampler, 254
financial concerns, 84, 89, 136, 173, 174, 261, 288
forms, 213–214, 216–217, 220, 260
founding, 30–32
ghost hunt at, 168, 222–224
haunting of, 167–168, 222–224, 228
insurance issue, 201–202
investigation protocols, 150
Kitt's living space, 77–79, 215–216, 218–219
Lifting Lodge, 287
membership, 63–64, 124–125
mission statement, 30, 32, 105
orb research, 43–46
Ouija boards banned, 137, 169–171
Psychic Sampler events, 179–180
psychomanteum project, 178–179, 214, 216–220
relocation and move, 221–222, 224–228
retail space, 85, 173–174, 214–215, 222–224
revenue streams, 77, 84, 173–174, 275–276
Rolling Hills Asylum clearing, 109–115

Index

SAGE classes, 245–246, 262, 264, 265
search for Solid Phantoms, 4–5, 276
Social Saucers, 81, 104–106, 129, 174, 242–243
Spiritual Community Gathering, 159, 161, 164, 165–166
Stevens's living space, 282–284, 286
triple-medium walkthrough, 145–154
vehicles, 39–40, 87, 89–91
Knotts, Don, 107
Krauss, Lieutenant, 209
KRI. *See* Kitt Research Initiative
Kukesh, Scott, 160, 163

Lane, Ed, 48–52, 58–59, 66, 78
Lavin, Melissa F., 272–273
Laz (ghost), 238
Lebanon College, 157
Leeuwenhock, Antonie van, 4
Leiden Observatory, 4
Lennon, John, 277
Leroux, Gaston, 173
Liberia, 198–199
Lifting Lodge, 287
Lin, Dacheng, 253, 293
Linnehan, Richard M., 101
Little Jack (ghost), 110, 115
lizard people, 176
LoConte, Dave, 163
Lofaso, Juliana, 62, 63, 159, 166, 226
Lofaso, Valerie
  alien encounters, 303
  assaulted by ghost, 148–150
  Bourdages and, 80–81, 84, 123, 124, 139–141, 151, 161, 206, 256
  CE-5 experience, 256–257
  Empath Support Group, 242, 288
  Esby and, 88–89, 108, 113, 256
  ghost hunting advice, 305

hearing voices, 110, 113, 256
Kitt and, 286–288
KRI membership, 63–65, 254–256
marriage and family, 62
paranormal interests, 62–63
personality, 61, 88–89
post-KRI, 288, 303–304
relationship with Dave, 275–276
school years, 61–62
as senior medium, 159, 275–276
soul lecture, 79–81
spirit guides, 253, 254, 257
SPRG, 63
Stevens and, 251–254, 256–258
table tipping experiment, 139–141
Tangled Web of Friends, 85
triple-medium walkthrough, 145–154
writings, 62, 63, 79, 83, 287
See also Kitt Research Initiative (KRI)
Louis XVI, 56–57
Lozi people, 299
Luthor, Lex, 180–183. *See also* Mazzaglia, Seth

MacDougall, Duncan, 80
Maclaine, Shirley, 277
Madison University, 177
Magnusson, Matthew, 242, 293
Maloney, Michael, 156, 158, 159–161, 162–164, 165, 166, 167
Marden, Kathleen, 26, 100
Marie Antoinette, 56
Markey, Dolly, 34–35, 155, 304
Markey, Paul, 35
Marriott, Lizzi, 184–186, 304
Massachusetts Bail Fund, 197
materialism, 142
"Materia Primoris" (Snow), 100

Index

Matthew Franklin Whittier in His Own Words (Sakellarios), 279
May, Ginny, 136–139, 143
Mayan Long Count calendar, 88
Mayer, Elizabeth, 72
Mazda Protege, 90–91
Mazzaglia, Seth, 183–187. *See also* Luthor, Lex
McDonough, Kat, 181, 184–186
McGehee, Marilyn, 69, 73
McIntosh College, 34–37, 155–156, 304
McLennan County Paranormal, 233
mediums
    church condemnation of, 92
    destigmatizing, 272–273
    Fox sisters, 92–96
    Holos experiment, 73–75
    inclusion of in ghost hunting, 136
    male, 179–180
    physical mediumship, 137
    triple-medium walkthrough, 145–154
    See also Boudages, Antje; Esby, Isabeau "Beau"; Lofaso, Valerie
Meetup, 16
memories, hypnosis-induced
    accuracy of, 58
    Kitt's, 48–50
    legal admissibility of, 53–55
    of past lives, 48–49, 277
    Stevens's, 50–52
mermaids, 175
Mesmer, Franz, 55–57, 92
mesmerism, 55–57
metaphysics, 80, 83, 198
Miles, Tiya, 234
Miller, Oritha, 35
Mitchell, Michael, 283, 286
Mitchell, Michelle, 283, 286
Mitchell Comics, 283, 286
Mojave Desert, 117–118

monetization
    of aliens, 104–106, 111
    of art collections, 299–300
    fictionalization of historical facts and, 234
    ghost hunting, 231
    of ghosts, 111, 115, 232–234
    of intellectual property, 103–104
    paratourism, 230, 231–234
Moody, Raymond, 178–179, 217, 218, 276
moon, 253
Mottin, Don, 59
Mount Washington, 155–156
Mozart, Wolfgang Amadeus, 56
murder, 182–187
Murphy, Jeremiah, 162–163
Muslims, belief in jinn, 269–270
Mutrie, Cullen, 160–161, 162–163, 165
Mysteries at the Museum (TV show), 104

Napoleon Bonaparte, 277
National Geographic (periodical), 131
near-death experiences, 178
New Age spiritualism, 186–187, 197, 291, 292
New Deal, 243
New England Society of Psychic Research, 27
New Hampshire
    Department of Transportation, 98
    distrust in government and social institutions, 2–3, 36–37, 157, 293
    Division of Historical Resources, 98
    ghost hunting groups, 33
    higher education, 53–55, 103–104, 155–157, 268–269
    historical highway markers, 97–98, 100–101, 104
    institutional closures, 36–37, 268

Index

paranormalism, 267
population, 156
public school enrollment decline, 269
Seacoast region, 5–6, 15
seasonal effects of climate change, 242
trust in paranormal believers, 2–3
"New Hampshire" (Frost), 13
New Hampshire Public Radio (NHPR), 100–101
NewsNation, 102, 252
New York Academy of Music, 94
New York Times (periodical), 101
normalization, 273

occultism, 70
Ocean Bank, 37
O. formigenes, 132
Ohio State Reformatory, 232, 233
Ohms, Jim, 189
Olson, Randy, 131, 293
orbs, 43–46
Orlando, Ernest, 203, 209, 210
Otto, Kate, 136–139, 143
Ouija boards, 137, 167n, 169–171

Page, Charles Grafton, 93–94, 135
palynology, 44
pancake, Kansas flatter than, 70
Paralights, 236
paranormal believers
  increase in, 144, 157, 292
  institutional distrust and, 91–92, 144, 233, 263
  paratourism, 230, 231–234
  post-institutional futures, 195–197, 297, 302–303
  research on, 70–72
  respecting, 272, 292
  skepticism towards, 71–72
  stigmatization of, 272–273
  trust in, 3–4
paranormal commerce, 231
paranormal investigators, 28–29
paranormalism, 252, 267, 289
paranormal research, 69–73, 75. *See also* Kitt Research Initiative (KRI)
parasites (spiritual), 222–224, 228
paratourism, 230, 231–234, 242
Parkes, Simon, 158
past-life regressions, 48–49, 276
Pasulka, Diana, 102–103, 197
Patton, George, 277
PBS, 157
perception, flaws in, 43–44
Pete Kitt's Service, 10
Peter (Esby's spirit guide), 110–111, 296
Pew Research Center, 71, 199
phrenology, 55
Pinkfloydia, 249
placebos, 57
plantation ghost tours, 234
Plato, 241
Poe, Edgar Allan, 279, 299
Popular Science (periodical), 131
Portsmouth Herald (periodical), 25
Portsmouth Naval Shipyard, 6
Portsmouth Regional Hospital, 163
post-institutional futures, 195–197, 198–200, 233, 297, 302–303
The Price is Right (TV show), 164
Priestley, Joseph, 130
professional scientists, 130–131
psychics, 254, 272–273. *See also* mediums
Psychic Sampler events, 179–180
psychodrama, 153
Psychological Medicine (periodical), 128–129
Psychomancy (Page), 93, 135

psychomanteum, 178–179, 214, 216–220, 260
psychophysics, 43–44, 70, 91, 207–208, 293
pterodactyls, 55
Putnam, Robert, 143
pysanky, 15

Radiating Electromagnetism Pod (Rem Pod), 236
Randi, James, 71–72
rapist, ghost of, 148–150, 152–154
raqi (healers), 270, 271
reincarnation, 276–282
religion
   belief in UFOs as, 102–103, 197
   decline of, 197
   dissemination of scientific knowledge, 130–131
   science and, 248–249
   See also churches; Spiritualism
R.E.M., 120
Rembrandt, 299
Rolling Hills Asylum, 29, 108–115
Roman Inquisition, 248
Roosevelt, Franklin, 243
Ross, Bill, 104
Roswell, NM, 118
Routledge, Clay, 198
Royal Geographical Society, 4
Runestone, 82, 287
Ryan, George, 143
Ryan, Jim, 211

Sagan, Carl, 276
SAGE (Spirituality, Alignment, Growth, and Empowerment) Method, 245–246, 262, 264, 265, 296–297
Saint Charles Borromeo Church, 36

Saint Denis Church, 36
Saint Michael the Archangel prayer, 155
Saint Patrick Church, 36
Sakellarios, Stephen, 275, 276–282
Salem, USS ("Sea Witch"), 230, 234–240
Salle des Sociétés Savantes, 141
The Sally Jessy Raphael Show, 28
Sanchez, Karina, 169, 293
Sanderson, Ivan Terrence, 247
Saturday Review (periodical), 71
Saturno, William, 88, 293
S. beyonceae, 249
Scarborough, Dorothy, 213
Schwartzman, Amelia Childs, 230–231
Scientific American (periodical), 131
scientific knowledge, 129–132, 248–249
scientific method, 129
Sci-Fi Channel, 29, 62
scrying, 178. *See also* psychomanteum
Seacoast Crime Stoppers, 156
Seacoast Drug Task Force, 160, 162
Seacoast Paranormal Research Group (SPRG)
   founding, 16–17
   ghost hunting, 29–30
   publicity, 25, 26
"Sea Witch" (warship). *See Salem*, USS
Seventh-Day Adventist Church, 97
Shackleford (ghost), 237, 238
Shakespeare, William, 9, 127
Shamballa Multi-Dimensional Healing, 175
Sharon (Rolling Hills Asylum employee), 109
Shealy, Norm, 69–70
shin-kicking, competitive, 27
Shor, Ronald E., 55, 58, 59, 104, 293

Shue, Erasmus, 93
Shue, Zona Heaster, 93
Silicon Valley, 103, 197–198
Sirius, 256
Skeptical Inquirer (periodical), 72
skeptics, 5
  alien encounters, 29, 208
  vs. believers, 58, 72, 170–171
  community of, 71–72, 95
  at Doyle's speech, 141, 142
  Esby's past as, 25, 82
  ghost hunting, 30, 235, 237
  ghosts and spirits, 139
  hypnosis, 58
  Kitt as, 175–176, 295
  orbs, 43
ski industry, 242
Skip (ghost), 237
Smith, Tom W., 157
Snow, Mark, 100
social capital, trust as the basic ingredient of, 2–3
Social Saucers, 81, 104–106, 129, 174, 242–243
Society for Promoting of Christian Knowledge, 130
sociology, 191–192, 272
Solid Phantoms, 4–5, 16, 43, 73, 75, 170, 246, 253, 272, 276, 295, 296, 298
Sorites paradox, 127, 268
soul, 79–80, 128, 129
spirit animals, 176–177
spirit guides, 18, 110–111, 245, 253, 254, 257, 287
spirit rapping, 137
spirits. *See* ghosts and spirits
"spiritual, but not religious" identifier, 199–200
Spiritualism, 92–96, 128, 141–142

Spiritualist Church (Boston), 137
spoon-bending, 284, 286
SPRG. *See* Seacoast Paranormal Research Group
Stallone, Sylvester, 164, 277
Stambovsky, Jeffrey, 228
Starbucks, 288
"Starlight" (The Free Design), 281
stars, 256
Stars/Time/Bubbles/Love (The Free Design), 281
Static Dome, 236
Stevens, Donna, 202, 209–211
Stevens, Michael, Jr. (Mike)
  alien abduction lecture, 81
  alien encounters, 24, 30–32, 41–42, 50–52, 105, 295
  arrests, 22–24, 210
  assault charges, 209–211
  Bourdages and, 285–286, 289
  CE-5 experience, 256–257
  as character in *Granite State Bigfoot*, 283
  conviction, 212
  divorce, 202
  Exeter UFO Festival, 282, 283–284, 302
  Granite Sky, 301
  highway marker petition, 26, 98, 100–101, 104, 282
  hypnosis, 50–52, 59
  jail term, 212, 226
  Kitt and, 26, 283–284, 289
  KRI living space, 282–284, 286
  KRI membership, 40–43
  Lofaso and, 251–254, 256–258
  murder attestation, 203, 205, 209
  post-KRI, 288–289, 301–303
  raccoon figurine, 168, 170–171
  release from jail, 242

Stevens, Michael, Jr. (Mike) *(continued)*
  Rolling Hills Asylum clearing, 109–115
  suicidal thoughts, 284, 286
  theft from KRI, 289
  Walton and, 208–209
  See also Kitt Research Initiative (KRI)
stigmatization, 272–273
Stine, Bill, 70, 91, 207
Stipe, Michael, 120
St. John International University, 156
Stratham Fitness Center and Holistic Spa, 17
suicide, 128, 284, 286–287
superstitions, 189–190
Sweeney, D. B., 206

table tipping, 137–141, 271
Tales from the Haunted South (Miles), 234
Tangled Web of Friends (Lofaso), 85
tarot, 181
telekinesis, 139
Texas State University, 70
The Atlantic Paranormal Society (TAPS), 28
Thomas, Cynthia Kile, 59
Thomas Institute of Hypnosis, 59
Thoreau, Henry David, 21
Tibbetts, Brittany, 161, 162, 165
TikTok, 231
tourism, 231. *See also* paratourism
trauma, 129, 153, 243, 271, 301, 303
Travel Channel, 104, 241
Trump, Donald, 292
trust
  as the basic ingredient of social capital, 2–3
  in Big Tech, 197–198, 269
  breakdown of, 2
  in government, 2, 3
  in institutions, 36–37, 103–104, 291–292
  table tipping and, 138–139
  See also distrust; institutions and institutional trust
Twain, Mark, 39, 97, 167n

UFOs
  belief in, 16, 102–103, 197
  Close Encounters of the Fifth Kind, 252
  Congress and, 102, 252
  Exeter Incident, 21–22, 282, 302
  Exeter UFO Festival, 282, 283–284, 302–303
  federal investigation of, 21
  French enthusiasts, 158, 159
  stigma for reporters of, 22
  See also alien encounters
ukiyo-e (Japanese woodblock prints), 299
UNH. *See* University of New Hampshire
University of California, Berkeley, 72
University of California, Santa Barbara, 270
University of Michigan, 143
University of New Hampshire (UNH)
  Betty and Barney Hill Collection, 104
  budget cuts, 103, 157, 191–192, 300
  Carsey School of Public Policy, 2–3, 103
  Conant Hall, 188
  Cooperative Extension, 157
  enrollment decline, 268, 300
  expansion of, 188
  federal-grant-funded programs cut, 268

Franklin Pierce Law Center,
  54–55, 103
Hersey House, 59
Horton Social Science Center, 188
institutional distrust, 191–193, 268
Kitt at, 15, 91, 188, 207, 287, 293
lack of faith in, 268–269
lecturers and programs cut, 268, 300
Mazzaglia at, 187
McConnell Hall, 188, 293
Museum of Art, 299–300
Paranormal Club, 187
psychology department, 70, 293
psychophysics, 43, 70
services provided by, 243
University of Pennsylvania, 94
US Congress, 102, 252
US Marine Corps, 35

Vatican Observatory, 102
vitalism, 55

Waco, TX, 232–233
Walking Elk (Esby's spirit guide), 18, 111, 245
Walton, Travis, 205–209, 211, 212, 216

Warhol, Andy, 299
Warren, Ed, 27–28, 137, 168, 304
Warren, Lorraine Moran, 27–28, 137, 168, 304
Washington, George, 97
whale songs, 169
White Mountain, 246
Whittier, Abby, 278–279, 280–282
Whittier, John Greenleaf, 277
Whittier, Mathew Franklin, 277–280
Williamson, Timothy, 127
Wilson, Grant, 28–29, 62, 108
Wilson, Samuel, 97
Woetzel, Dave, 248
Wöhler, Friedrich, 55
Woods Devil, 246, 258. *See also* Bigfoot
World War II, 9–10
Worsted Church, 97
Wynands, Anthony, 186

The X-Files (TV show), 100

YouGov, 247
Yucca Men, 118

Zach (paratourist), 237
Zerbe, Karl, 299

MATTHEW HONGOLTZ-HETLING is a freelance journalist specializing in narrative features and investigative reporting. He has been named a finalist for the Pulitzer Prize, won a George Polk Award, and been voted Journalist of the Year by the Maine Press Association, among numerous other honors. He has written for *Foreign Policy*, the *New Republic*, *USA Today*, *Popular Science*, *Atavist Magazine*, the Pulitzer Center on Crisis Reporting, the Associated Press, and elsewhere. He is the author of two prior books, *A Libertarian Walks Into a Bear* and *If It Sounds Like a Quack*.